Cambridge Tracts in Mathematics and Mathematical Physics

No. 61

GENERALIZED CLIFFORD PARALLELISM

GENERALIZED CLIFFORD PARALLELISM

J. A. TYRRELL
Reader in Mathematics,
King's College London

AND

J. G. SEMPLE
Professor Emeritus of Mathematics,
King's College London

CAMBRIDGE
AT THE UNIVERSITY PRESS
1971

Published by the Syndics of the Cambridge University Press
Bentley House, 200 Euston Road, London N.W.1
American Branch: 32 East 57th Street, New York, N.Y.10022

Library of Congress Catalogue Card Number: 74–134625

ISBN: 0 521 08042 8

Printed in Great Britain
at the University Printing House, Cambridge
(Brooke Crutchley, University Printer)

CONTENTS

Preface — *page* vii

1 Introduction — 1

2 Preliminaries of geometry in S_{2n-1} — 10

3 Clifford parallel spaces and Clifford reguli — 20

4 Linear systems of Clifford parallels — 34

5 Geometrical constructions — 58

6 The T-representation — 82

7 Half-Grassmannians — 110

Appendix. The Hurwitz–Radon matrix equations — 127

References — 138

Index — 140

CONTENTS

Preface *page* vii

1 Introduction 1

2 Preliminaries of geometry in S_{2n-1} 10

3 Clifford parallel spaces and Clifford reguli 20

4 Linear systems of Clifford parallels 34

5 Geometrical constructions 58

6 The T-representation 82

7 Half-Grassmannians 110

Appendix. The Hurwitz–Radon matrix equations 127

References 138

Index 140

PREFACE

The subject of parallelism in elliptic non-Euclidean space is one that has hitherto received little attention in works on geometry. In fact, although Clifford's original theory of parallel lines in three-dimensional elliptic space goes back nearly a hundred years, the generalization to higher space, due to Y.-C. Wong, was not devised until comparatively recently. On reading Wong's work (which is predominantly algebraic and mainly confined to geometry over the real field) it occurred to the authors that the subject could more naturally be developed in the wider context of complex projective geometry and the present tract was conceived with this object in mind.

The general plan of the book, together with some points of history and motivation, is the subject of Chapter 1 and need not be anticipated here. As to the style of writing, we have tried to develop the earlier Chapters in a careful and rigorous manner, paying full attention to the details of the arguments; in the later Chapters, however, we have felt free to use the less formal language of the classical geometers and to omit some of the more routine proofs and calculations.

We wish to thank Dr D. Kirby of the University of Southampton for the meticulous care with which he read our page proofs and for his numerous corrections and emendations; and our warm thanks are also due to the officers of the Cambridge University Press for their unfailing courtesy and assistance in the preparation of this volume.

King's College London J.A.T.
June 1970 J.G.S.

CHAPTER 1

INTRODUCTION

1. Parallelism in the plane

When the study of plane Euclidean geometry is being developed by coordinate methods, there are two normally accepted procedures for dealing with the phenomenon of parallelism. In the first of these procedures, we regard the Euclidean plane, effectively, as a real projective plane with one line removed—a line which we denote by l_∞; we can then regard the projective plane, conversely, as an *extended Euclidean plane*; and thereafter we define parallel lines as lines whose closures in the extended Euclidean plane meet on l_∞. By the second procedure we first introduce a distance function and the concept of perpendicularity, showing that the shortest distance from a point to a line is the perpendicular distance; and we then say that two lines are parallel if each point of either is at the same perpendicular distance from the other.

When we come to study non-Euclidean geometry, it is natural to investigate the possibility of generalizing one or other of the above approaches to parallelism. A non-Euclidean plane, as we may recall, is basically a real projective plane in which there is singled out a particular polarity ω—the *absolute polarity*; that is to say, in a chosen system of homogeneous coordinates (x_0, x_1, x_2) in the plane, there is singled out a bilinear relation

$$\sum_{i=0}^{2} \sum_{j=0}^{2} a_{ij} x_i x'_j = 0 \tag{1.1}$$

such that the matrix $A = (a_{ij})$ is real, non-singular and symmetric; and two points (x_0, x_1, x_2) and (x'_0, x'_1, x'_2) are said to be *conjugate* with respect to the polarity ω if their coordinates satisfy (1.1). The points of the plane that are conjugate to themselves are those of the (non-singular) *absolute conic* Ω whose equation is

$$\sum_{i=0}^{2} \sum_{j=0}^{2} a_{ij} x_i x_j = 0;$$

and the non-Euclidean plane is said to be *hyperbolic* or *elliptic* according as Ω does or does not contain any real points; or, as we usually say, according as Ω is a *real* or a *virtual*† conic.

In the hyperbolic case Ω disconnects the projective plane into two regions, one homeomorphic with a disk and the other with a Möbius strip; but it is customary to think of the hyperbolic plane as properly consisting only of the points of the former region, those of the latter region being called *ultra-infinite points* and those of Ω *points at infinity*. The elliptic plane, on the other hand, consists of *all* the points of the underlying projective plane.

The *non-Euclidean distance* $d(P,Q)$ between points P and Q of a non-Euclidean plane is defined in terms of the cross-ratio $\{P,Q;Q',P'\}$, where P' and Q' are the points of the line PQ that are conjugate to P and Q respectively in the polarity ω. The usual definition is

$$\left.\begin{array}{r}\cosh^2(d(P,Q))\\ \cos^2(d(P,Q))\end{array}\right\} = \{P,Q;Q',P'\},$$

the hyperbolic or trigonometric cosine being taken according as the plane is hyperbolic or elliptic.‡ The use of the term 'point at infinity' for a point of Ω (in the hyperbolic case) is justified by the fact that $d(P,Q)$ tends to infinity if P is kept fixed while Q is made to approach a point of Ω.

After these preliminaries we may now consider possible generalizations to non-Euclidean geometry of the two ways of introducing parallel lines in the Euclidean plane. By analogy with the first of these ways we may define two lines of a non-Euclidean plane to be parallel if they meet 'at infinity', i.e. on Ω. In the hyperbolic case this definition makes sense, and its consequences have been fully worked out in treatises on non-Euclidean geometry (cf., for example, Coxeter [**6**] which contains an extensive bibliography). In the elliptic case, however, the definition is empty because Ω contains no real points.

† Since virtual conics contain no real points, they do not exist (as loci) in the same sense as real conics. They are associated, however, with well-defined real polarities and thereby acquire a conventional 'existence' as conics in the real projective plane.

‡ The formulae given above do not define $d(P, Q)$ unambiguously in either case. For the conventions necessary to their interpretation we refer the reader to the account given in Coxeter [**6**].

As regards the possible generalization of the 'equidistance' definition of parallel lines, we note first that *perpendicular lines* in either kind of non-Euclidean geometry are defined to be lines which are conjugate with respect to the absolute polarity ω, and that the shortest distance from a point to a line is the perpendicular distance. We then find, however, that no two distinct lines can ever have the 'equidistance' property; and, in fact, the locus of a point which is at a fixed perpendicular distance from a given line – *an equidistant curve* – is (algebraically) a conic touching Ω at its intersections with the line. There is, accordingly, no analogue, in either kind of non-Euclidean plane, of the equidistance definition of parallelism in the Euclidean plane.

2. Parallel lines in 3-dimensional space

Consider now the corresponding problems in 3-dimensional space. Parallel lines in Euclidean 3-dimensional space can be introduced in either of two ways, either as lines that meet on a conveniently introduced 'plane at infinity' or as lines that are everywhere equidistant; and we look again at the possibility of adapting one or other of these definitions to the circumstances of 3-dimensional non-Euclidean geometry.

A 3-dimensional non-Euclidean space is a 3-dimensional real projective space in which there is singled out a certain fixed (non-singular) quadric polarity ω—the *absolute polarity*—given in a chosen coordinate system by an equation

$$\sum_{i=0}^{3}\sum_{j=0}^{3} a_{ij}x_i x'_j = 0,$$

where $A = (a_{ij})$ is a non-singular real symmetric matrix. We again distinguish different kinds of non-Euclidean space according to the nature of the non-singular *absolute quadric* Ω whose equation is

$$\sum_{i=0}^{3}\sum_{j=0}^{3} a_{ij}x_i x_j = 0.$$

Specifically, the space is *elliptic* if Ω is a virtual quadric (containing no real points) and *hyperbolic* if Ω is a real quadric of the kind that is homeomorphic with an ordinary 2-dimensional sphere; and in the latter case we again distinguish (as in the plane) between ordinary, infinite and ultra-infinite points.† *Non-Euclidean distance* is defined, in either kind of space, by the same formulae as in the plane; and two lines p and q are defined to be *perpendicular* if they are conjugate with respect to ω, i.e. if each meets the polar line of the other. It is easy to verify, then, that the shortest distance from a point to a line is the perpendicular distance.

As regards the possible definitions of parallel lines, the first possibility is to define two lines as being parallel if they meet on Ω; and, just as in the plane, this definition makes sense in the hyperbolic but not in the elliptic case. With the equidistance definition, however, an exciting new possibility arises, of which there was no analogue in the plane. The locus ϕ of points at a fixed perpendicular distance from a given line l – an *equidistant surface* – turns out to be a quadric surface which (algebraically) touches Ω at its two intersections with l; and there is now the possibility that this quadric ϕ may be of the type that contains real generating lines. It then turns out that this possibility is realized if (and only if) the non-Euclidean space is elliptic. To recapitulate formally:

In 3-dimensional elliptic space, the locus of points at a fixed perpendicular distance from a given line l is a real quadric surface ϕ with (two systems of) real generating lines.

Any line of either generating system on ϕ is therefore parallel to l in the 'equidistance' sense, and it is said to be *Clifford parallel*‡ to l. Plainly, through any general point P (of elliptic space) there pass two lines that are Clifford parallel to a given line l, one generator of each system on the quadric such as ϕ that contains P; and it is possible to introduce conventions whereby one of these lines can be said to be *left-parallel* to l and the other *right-parallel* to l. Left-parallelism and right-parallelism each turn out to be

† There is, of course, another kind of real non-singular quadric in real projective space, namely the kind – homeomorphic with a torus – which possesses two systems of real generating lines; but for various reasons that need not detain us here, this kind of quadric does not give rise to an acceptable kind of non-Euclidean geometry.

‡ In the older literature, Clifford parallel lines were sometimes called *paratactics*.

transitive relations, though Clifford parallelism without qualification is not. An important consequence of the above remarks is that there exist *fibrations* of elliptic 3-space by systems of mutually left- (or right-) parallel lines, where by a fibration we mean a system of which exactly one line passes through each point of space (with no exceptions).

3. Clifford's original definition of parallelism

We now refer to another way of approaching the concept of Clifford parallelism, more in the spirit of Clifford's original discovery (cf. Clifford [**5**]). Suppose that we are given a point P and a line q in elliptic 3-space, and let q' be the polar line of q in the absolute polarity ω. Then the transversal line from P to q and q' is the perpendicular from P to q, and the shortest distance from P to q is measured along this perpendicular. Now suppose that P moves along a line p whose polar line is p'. It can be shown then that the distance from P to q varies continuously as P moves along p, and it attains its maximum and minimum values when P lies on one of the two transversal lines t and t' that can be drawn to meet p, p', q and q'. These lines t and t' are the common perpendicular transversals of p and q; and, provided p and q are in sufficiently general position, there are no others. But, as Clifford noticed, there is a case of exception in which p, p', q and q' belong to the same regulus of generators of a ruled quadric R. In this case p and q possess an infinite number of common perpendicular transversals – the lines of the regulus on R complementary to that containing p, q, p' and q' – and the distances between p and q along these transversals are all equal. These observations lead to the following projective characterization of Clifford parallel lines:

Clifford's definition of parallelism: *Two lines p and q are (Clifford) parallel if (and only if) they and their polar lines with respect to ω are four generators of one regulus on a ruled quadric R.*

A quadric such as R, i.e. one which contains two Clifford parallel lines and their polars, possesses the following two (characteristic) properties:

(*a*) Any two non-intersecting generators of R are Clifford parallel; more precisely, the generators of one regulus on R are all mutually left-parallel, while those of the other are mutually right-parallel; and

(*b*) R is autopolar with respect to ω, i.e. it contains the polar line of every one of its generators.

A quadric with properties (*a*) and (*b*) will be called a *Clifford quadric*,† and each of its two reguli will be called a *Clifford regulus*. In the higher dimensional analogues of Clifford parallelism which we shall be studying, it will be found that the appropriate analogue of a Clifford regulus plays a fundamental role.

4. Isoclinic planes in Euclidean space E_4

Further insight into Clifford parallelism may be gained by considering the problem, in 4-dimensional Euclidean space E_4, of defining the angle between two planes (2-dimensional subspaces of E_4).

If π_1 and π_2 are two non-orthogonal planes through a point O of E_4, then one may consider a variable vector v through O in π_1 and calculate the angle between v and its orthogonal projection on π_2. This angle, as it turns out, varies with v in general, achieving its maximum and minimum values when v lies in one or other of two mutually perpendicular directions in π_1. Thus from this approach it is not clear how to define the angle between π_1 and π_2. (For the answers to this and related questions the reader may consult Forsyth [**10**].) There is a case of exception, however, that can and does occur, in which the angle θ between v and its projection on π_2 is independent of the choice of v in π_1. Here there is no doubt as to how to define the angle between π_1 and π_2; and we say that π_1 and π_2 are *isoclinic* at an angle θ.

It is not our purpose to investigate the properties of isoclinic planes in any detail; this has been done elsewhere (see, for example, Forsyth [**10**] and Manning [**19**]). We only wish to point out that Euclidean space E_4 may be thought of as acquiring its Euclidean structure in terms of a 'solid at infinity' together with

† We should point out that Coxeter [**6**] uses the term 'Clifford surface' to describe any quadric which possesses property (*a*) but not necessarily (*b*).

a virtual quadric Ω in this solid; and the solid at infinity has then the structure of an elliptic 3-dimensional space. It may now be verified easily that *two planes of E_4 through O are isoclinic if and only if they meet the elliptic 3-dimensional space at infinity in Clifford parallel lines.*

To mention only one consequence of this, consider a fibration of the solid at infinity by a system of mutually left-parallel lines, say; and join these by planes to a fixed origin O in E_4. Exactly one of these planes will pass through any point of E_4 other than O. Thus if we consider a 3-dimensional sphere S^3 in E_4 with centre O, we see that the planes in question will cut S^3 in a family of non-intersecting great circles such that exactly one passes through any point of S^3. The decomposition space of this family of circles turns out to be a 2-dimensional sphere S^2; and what we have, in effect, is the well-known Hopf fibering of S^3 by 1-spheres over S^2. In terms of this construction one can further establish the existence of three mutually orthogonal fields of unit tangent vectors on S_3, thus demonstrating its parallelizability.

5. Generalization of Clifford parallelism

A $(2n-1)$-dimensional elliptic space EL_{2n-1} is a $(2n-1)$-dimensional real projective space in which a certain virtual quadric is assigned the role of absolute quadric Ω and the associated real non-singular polarity is the absolute polarity ω. An $(n-1)$-dimensional subspace Π of EL_{2n-1} has a polar space Π' with respect to ω, also of dimension $n-1$, and a line is said to be perpendicular to Π if it meets Π'. If we define distance in the same way as for the elliptic plane, we find that the shortest distance to Π from a point is the perpendicular distance, and we may make the following definition:

Two $(n-1)$-dimensional subspaces Π and Π_1 of EL_{2n-1} are to be called *Clifford parallel* if the perpendicular distance to Π_1 from any point P of Π is independent of the choice of P in Π.

This is no more than a straightforward generalization of the original concept of Clifford parallelism, and it is therefore remarkable that it was not systematically studied and developed until recently. The recent development, due to Wong [**27**], is coupled with a parallel study of isoclinic n-dimensional subspaces of Euclidean space E_{2n}, the connection with Clifford parallelism being analogous to that which we have described above (§ 4) for $n = 2$. Wong's development is primarily directed to the discovery of systems – more particularly maximal systems – of mutually Clifford parallel $[n-1]$'s† in EL_{2n-1}; and we may summarize his findings briefly as follows:

(i) For all values of n (> 1) there exist 1-dimensional systems of mutually Clifford parallel $[n-1]$'s in EL_{2n-1}. Important among such systems are those which Wong calls *additive sets*, and these have properties very similar to those of Clifford reguli in the case $n = 2$.

(ii) For given n there exists an r-dimensional system of mutually Clifford parallel $[n-1]$'s in EL_{2n-1} if and only if there exists a set of $r-1$ skew-symmetric real orthogonal matrices of order n which anticommute by pairs. The algebraic problem so arising is that of solving the well-known Hurwitz–Radon matrix equations, and the pairs of values (n, r) for which a solution exists are known‡ (cf. the Appendix to this book).

(iii) Except when $n = 2, 4$ or 8, there exist no n-dimensional systems of mutually Clifford parallel $[n-1]$'s in EL_{2n-1}, and consequently no fibrations of EL_{2n-1} based on such systems; but when $n = 2$, 4 or 8, such fibrations (or *space-filling systems*) do exist. The fibrations for $n = 4$ and $n = 8$ correspond, by a construction similar to that described above for $n = 2$, to the Hopf fiberings of the 7-sphere S^7 by S^3's over S^4 and of the 15-sphere S^{15} by S^7's over S^8.

Since Wong's treatment of Clifford parallelism, and the formulation of his results, were predominantly algebraic and limited to geometry over the real field, it seemed to the present authors that the subject invited an alternative development

† The symbol $[r]$ will be used in this book to denote an r-dimensional (flat) space.

‡ When $n = 3$, the maximum possible value of r is 1; so that in this, the next simplest case of Clifford parallelism after the classical case, the phenomenon is relatively uninteresting. This may partly explain why the problem of generalizing Clifford parallelism did not attract attention for such a long time.

within the general framework of complex projective geometry; and, in particular, it seemed highly desirable to seek some insight into the geometric construction and properties of systems of mutually Clifford parallel spaces. In this book we give such a development, essentially independent of that given by Wong, and consisting basically of a chapter in the projective geometry of complex projective space S_{2n-1} relative to a non-singular quadric primal Ω_{2n-2}. After some preliminaries (Chapter 2), we introduce the notion of Clifford parallel spaces with respect to Ω, using a straightforward generalization of Clifford's own definition (§ 3), and we rediscover Wong's additive sets (which we call Clifford reguli) and develop their properties (Chapter 3). This is followed by a mainly algebraic exposition of what we call *linear* systems of Clifford parallels (Chapter 4), analogous to, but rather more general than the systems discovered by Wong. The next Chapter is devoted to geometrical constructions for systems of Clifford parallels, particular attention being paid to the two most interesting cases, $n = 4$ and $n = 8$. The remainder of the book (Chapters 6 and 7) is concerned with representation theory; and this includes in particular a detailed account of a new sequence of algebraic varieties – here called T-models – which serve as birational maps of a specially significant kind of the pairs of polar $[n-1]$'s with respect to a quadric Ω in S_{2n-1}. Among the applications of our representation theory we encounter in Chapter 7 an apparently new construction for, and extended interpretation of, that well-known geometrical curiosity, the Study triality correspondence between the points and the two types of generating solids of a quadric Ω in S_7.

CHAPTER 2

PRELIMINARIES OF GEOMETRY IN S_{2n-1}

In this book we shall be concerned throughout, except when the contrary is stated, with geometry in a complex $(2n-1)$-dimensional projective space S_{2n-1} relative to a non-singular 'absolute' quadric primal Ω. We devote the present chapter, therefore, to recording some preliminary material about the appropriate coordinate systems, the properties of a quadric Ω, the definition and properties of the well-known type of algebraic variety which we call a *regulus* of $(n-1)$-dimensional spaces of S_{2n-1}, and some definitions and details relating to Grassmannian and Veronesean varieties.

1. Coordinate system in S_{2n-1}

In S_{2n-1} we shall generally use a system of allowable homogeneous coordinates

$$(x_0, \ldots, x_{n-1}, y_0, \ldots, y_{n-1}) \tag{1.1}$$

based on a pair of skew $(n-1)$-dimensional reference spaces X, Y such that X contains the reference points $X_0, \ldots, X_{n-1}$ while Y contains $Y_0, \ldots, Y_{n-1}$. We shall write x (resp. y) for the column vector of the x_i (resp. y_i) and shall refer to (1.1) as the point (x, y). We note then that

(i) any $[n-1]$ of S_{2n-1} which is skew to the reference space Y has a matrix equation of the form

$$y = Ax, \tag{1.2}$$

where A is a uniquely defined $n \times n$ matrix of constants, and

(ii) the same $[n-1]$ is skew also to X if and only if $|A| \neq 0$. Two $[n-1]$'s given by $y = A_1 x$ and $y = A_2 x$ are skew to one another if and only if $|A_1 - A_2| \neq 0$.

2. The non-singular quadric Ω

We shall generally choose the coordinate system in such a way that the equation of Ω reduces to one or other of two standard forms which we now consider separately.

The first standard form $x^Tx+y^Ty = 0$. For this form, in addition to a proper choice of the unit point, the simplex of reference must be self-polar for Ω; and, in particular, the reference spaces X and Y must be polars of each other for Ω. Conversely, moreover, it should be noted – as may easily be verified – that any pair of skew mutually polar $[n-1]$'s may be chosen as the reference spaces X, Y of a coordinate system in which Ω has the equation $x^Tx+y^Ty = 0$. The condition that a pair of polar $[n-1]$'s should be skew means that neither of them touches Ω, i.e. meets Ω in a singular quadric.

When Ω has the equation $x^Tx+y^Ty = 0$ and Π is the $[n-1]$ with equation $y = Ax$, then it is easily seen that

(i) the section of Ω by Π is non-singular if and only if $|I+A^TA| \neq 0$, where I is the $n\times n$ unit matrix,

(ii) if Π does not meet X, so that $|A| \neq 0$, then the polar space Π' of Π has the matrix equation $y = -A^{-1T}x$, and

(iii) Π lies on Ω (i.e. is self-polar) if and only if $A^TA = -I$.

The second standard form $x^Ty = 0$. For this form, besides the choice of unit point, the reference spaces X and Y must be a pair of skew generating spaces of Ω; and, conversely, any such pair of skew generators of Ω can be taken to be the reference spaces X, Y of a coordinate system in which Ω has the equation $x^Ty = 0$.

When Ω has the equation $x^Ty = 0$ and Π is the $[n-1]$ given by $y = Ax$, then

(i) the section of Ω by Π is non-singular if and only if

$$|A+A^T| \neq 0,$$

(ii) the polar space Π' of Π is given by $y = -A^Tx$, and

(iii) Π is a generator of Ω if and only if A is skew-symmetric.

The equation $x^Ty = 0$ can also be interpreted as that of a correlation between the spaces X and Y with the property that points P and Q in X and Y respectively correspond in the correlation if and only if their join PQ lies on Ω. Thus Ω induces in this way what we may call a *natural* correlation κ between any skew pair of its generating spaces.

We may recall here the following well-known properties of generators of Ω:

(*a*) Ω possesses two distinct systems of $(n-1)$-dimensional generating spaces (which we shall call the α-spaces and the β-spaces), each system being of dimension $\frac{1}{2}n(n-1)$.

(*b*) If n is even, two generators of the same system do not meet in general but may meet in a space of an odd number of dimensions, while two generators of opposite systems meet always in a point or in a space of an even number of dimensions.

(*c*) If n is odd, two generators of the same system meet either in a point or in a space of an even number of dimensions, while two of opposite systems do not in general meet but may meet in a space of an odd number of dimensions.

3. Reguli in S_{2n-1}

In the developments to follow we shall be much concerned with the type of algebraic variety R of S_{2n-1} that is generated by the lines joining pairs of points which correspond in a non-singular collineation between two skew $[n-1]$'s. Any such R is readily identified as a Segre product of a line by an $(n-1)$-dimensional space, and it is known, as such, to be a non-singular variety R_n^n normal in S_{2n-1}. The product structure shows that R is simply generated not only by its ∞^{n-1} lines, which are called its *directrices*, but also by ∞^1 mutually skew $[n-1]$'s – unisecant to the directrices – which are called its *generators*. A variety such as R, particularly when it is regarded as the aggregate of its generators, will be called a *regulus* (of $[n-1]$'s of S_{2n-1}). There exist, of course, specialized and degenerate types of the aggregate in question; but we shall find it convenient throughout this book to reserve the term 'regulus of $[n-1]$'s' to mean always an aggregate which conforms precisely to the general definition we have given.

Analytically, if we define R by means of the collineation

$$\rho y = Ax \quad (|A| \neq 0)$$

between the reference spaces X and Y, we find that the generators of R are given, for different values† of the scalar parameter λ, by the equation

$$y = \lambda A x.$$

With no loss of generality (i.e. by a transformation of the coordinates x_i) we can take A to be the $n \times n$ unit matrix, in which case the generators of R are given by the equation

$$y = \lambda x,$$

and the equations of R are

$$\frac{x_0}{y_0} = \frac{x_1}{y_1} = \ldots = \frac{x_{n-1}}{y_{n-1}}.$$

We note, in particular, that R is completely defined as an intersection of quadrics.

Among the elementary results about reguli we record briefly the following.

PROPOSITION 3.1. *The directrix lines of a regulus R cut collinear fields on its generators.*

This means, in effect, that the points of any set S_1 in a generator Π_1 are carried along the directrices passing through them into the points of a collinear set S_2 in any other generator Π_2. We shall express this connection frequently by saying that S_2 *lies over* S_1 *(or* S_1 *over* S_2*) in* R.

PROPOSITION 3.2. *The transversal lines of three mutually skew $[n-1]$'s of S_{2n-1} are the directrices of a regulus which contains the three $[n-1]$'s as generators.*

For the correspondence between two of the $[n-1]$'s determined by transversals drawn to them from points of the third is a collineation.

COROLLARY 3.2.1. *There is a unique regulus R containing three mutually skew $[n-1]$'s of S_{2n-1}.*

† The space Y corresponds to the improper value ∞ of λ. For any letter denoting a parameter, it will usually be clear from the context whether ∞ is to be included in its range of values.

PROPOSITION 3.3. *There exist* ∞^{2n-4} *solids (3-dimensional flat spaces) each of which meets every generator of a given regulus* R *in a line (these lines all lying over any one of themselves and generating a quadric surface); and precisely one of these solids passes through any point of* S_{2n-1} *that does not lie on* R.

The solids in question will be referred to as the *quadric-solids* of R.

We leave the proof of Prop. 3.3 as an exercise for the reader.

PROPOSITION 3.4. *Let* S_{2p-1} *and* S_{2q-1} $(p+q=n)$ *be two skew spaces which span* S_{2n-1}*; and let* R_1 *be a regulus of* $[p-1]$*'s of* S_{2p-1}*, while* R_2 *is a regulus of* $[q-1]$*'s of* S_{2q-1}*. Further, let there be given a homographic correspondence between the generators of* R_1 *and those of* R_2*. Then the* $[n-1]$*'s which join corresponding generators of* R_1 *and* R_2 *are the generators of a regulus of* $[n-1]$*'s in* S_{2n-1}.

Here again the proof may be left to the reader. A similar result holds for three or more homographically related reguli in mutually skew spaces which span S_{2n-1}.

4. The matrix cross-ratio theorem

We now establish a basic algebraic criterion for four $[n-1]$'s of S_{2n-1} to belong to a regulus.

THEOREM 4.1. *If* $A_1, \ldots, A_4$ *are four* $n \times n$ *matrices for which all the differences* $A_i - A_j$ $(i \neq j)$ *are non-singular, and if* Π_i *is the* $[n-1]$ *of* S_{2n-1} *with equation* $y = A_i x$ $(i = 1, \ldots, 4)$*, then* $\Pi_1, \ldots, \Pi_4$ *belong to a regulus* R *if and only if there exists a scalar* c $(\neq 0, 1)$ *such that*

$$(A_1 - A_3)(A_1 - A_4)^{-1}(A_2 - A_4)(A_2 - A_3)^{-1} = cI, \qquad (4.1)$$

where I *is the* $n \times n$ *unit matrix. If the regulus* R *is thought of abstractly as a projective line whose points are the generators of* R*, then* c *is the cross-ratio* $\{\Pi_1, \Pi_2; \Pi_3, \Pi_4\}$.

Proof. The non-singularity of the matrices $A_i - A_j$ $(i \neq j)$ means simply that the four spaces $\Pi_1, \ldots, \Pi_4$ are mutually skew.

Consider an arbitrary regulus R containing Π_1 and Π_2. Then R is defined by some non-singular collineation between the points $(\xi_1, A_1\xi_1)$ of Π_1 and the points $(\xi_2, A_2\xi_2)$ of Π_2, say

$$\rho\xi_2 = T\xi_1 \quad (|T| \neq 0),$$

and the general point of the variety R is of the form

$$(\xi_1 - \lambda T\xi_1, A_1\xi_1 - \lambda A_2 T\xi_1).$$

The directrices of R are obtained by fixing ξ_1 and allowing λ to vary, while the generators arise from fixing λ and allowing ξ_1 to vary. We therefore get the general generator by eliminating the vector ξ_1 between the two equations

$$x = \xi_1 - \lambda T\xi_1 \quad \text{and} \quad y = A_1\xi_1 - \lambda A_2 T\xi_1.$$

This yields
$$y = (A_1 - \lambda A_2 T)(I - \lambda T)^{-1}x,$$

and we note that $\lambda = 0$ gives Π_1, while $\lambda = \infty$ gives Π_2.

If we now require Π_3 and Π_4 to belong to R, we must have

$$A_i = (A_1 - \lambda_i A_2 T)(I - \lambda_i T)^{-1} \quad (i = 3, 4) \qquad (4.2)$$

for suitable scalars λ_3 and λ_4; and since $\Pi_1, \ldots, \Pi_4$ are distinct, λ_3 and λ_4 must be distinct and neither of them equal to 0 or ∞. A rearrangement of (4.2) gives

$$T = \lambda_i^{-1}(A_2 - A_i)^{-1}(A_1 - A_i) \quad (i = 3, 4)$$

so that, equating these two expressions for T, we have

$$\lambda_3^{-1}(A_2 - A_3)^{-1}(A_1 - A_3) = \lambda_4^{-1}(A_2 - A_4)^{-1}(A_1 - A_4).$$

This equation on rearrangement gives (4.1) with $c = \lambda_3/\lambda_4$, and since λ_3 and λ_4 are distinct non-zero constants we have $c \neq 0, 1$. Since the algebra is reversible, the first part of the theorem is thereby proved.

Further, since $\Pi_1, \ldots, \Pi_4$ are the members of R given respectively by the values $0, \infty, \lambda_3, \lambda_4$ of λ, we have

$$\{\Pi_1, \Pi_2; \Pi_3, \Pi_4\} = \{0, \infty; \lambda_3, \lambda_4\} = \lambda_3/\lambda_4 = c;$$

and this completes the proof of the theorem.

It is natural to denote the left-hand side of (4.1) by

$$\{A_1, A_2; A_3, A_4\}$$

and to call it the *cross-ratio of the four matrices*. It is then implicit in Theorem 4.1 that if $\{A_1, A_2; A_3, A_4\} = cI$, then – under permutations of $A_1, \ldots, A_4$ – the cross-ratio behaves like the cross-ratio of four numbers. For example:

$$\{A_2, A_1; A_3, A_4\} = c^{-1}I$$

and

$$\{A_1, A_3; A_2, A_4\} = (1-c)\,I.$$

Formal algebraic proofs of these results involve some skilful matrix manipulations.

Ex. 1. Let

$$A_0 = \begin{pmatrix} \alpha_0 & \beta_0 \\ \gamma_0 & \delta_0 \end{pmatrix} \quad \text{and} \quad A_1 = \begin{pmatrix} \alpha_1 & \beta_1 \\ \gamma_1 & \delta_1 \end{pmatrix}$$

be two given matrices of order two with the property that the difference between any two of A_0, A_0^T, A_1, A_1^T is non-singular. For a general value of the scalar λ, show that the equation

$$\{A_1, A_0; A_0^T, X\} = \lambda I$$

has a unique solution for X, namely,

$$X = A(\lambda) \equiv (\lambda A_1 K - A_0)(\lambda K - I)^{-1},$$

where $K = (A_1 - A_0^T)^{-1}(A_0 - A_0^T)$. $A(\lambda)$ is equal to A_0, A_1 and A_0^T respectively when $\lambda = 0$, ∞ and 1. Show also that $A(\lambda)$ is equal to A_1^T when λ takes the value λ_1 given by

$$(\beta_0 - \gamma_0)(\beta_1 - \gamma_1)\lambda_1 = \det(A_1 - A_0^T).$$

It follows that *two general matrices of order two, together with their transposes, satisfy the condition of the matrix cross-ratio theorem*. The same result is not true for matrices of higher order.

Ex. 2. With the notation of Ex. 1, suppose also that the eigenvalues of A_0 and A_1 are (ψ_0, ψ_0') and (ψ_1, ψ_1') respectively. Show then that the eigenvalues of $A(\lambda)$ are the roots of the equation

$$(\psi - \psi_0)(\psi - \psi_0') + \theta(\psi - \psi_1)(\psi - \psi_1') = 0,$$

where

$$\theta = \frac{\beta_0 - \gamma_0}{\beta_1 - \gamma_1} \cdot \frac{\lambda - \lambda^2}{\lambda - \lambda_1}.$$

It follows that, if A_0 and A_1 have no common eigenvalue, then the eigenvalues of $A(\lambda)$ are pairs which correspond in a certain involution. On the other hand if, say, $\psi_1 = \psi_0$, then ψ_0 is also an eigenvalue of $A(\lambda)$.

5. Grassmannian varieties

In this and the following section we shall refer briefly to some of the properties of two types of algebraic variety that will feature largely in the subsequent development. The first of these is the *Grassmannian variety* $G(k, m)$ which is the standard unexceptional model of all the k-dimensional subspaces S_k of a projective space S_m. For detailed accounts of this variety we refer the reader in particular to the memoir [**23**] of Severi and to Vol. 1, Ch. 7 and Vol. 2, Ch. 14 of Hodge and Pedoe [**13**].

We shall assume, then, that the reader is familiar with the procedure of representing an S_k of S_m by a point P of S_M, where

$$M = \binom{m+1}{k+1} - 1,$$

the homogeneous coordinates of P being a systematic arrangement of the determinants of order $k+1$ extracted from the $(k+1) \times (m+1)$ matrix of the coordinates of any $k+1$ linearly independent points of the S_k. The locus of P as the S_k varies is the Grassmannian $G(k, m)$ of the S_k's of S_m. The dual Grassmann coordinates of the S_k, obtained in a similar way from the coordinate matrix of any $m-k$ linearly independent primes through S_k, turn out to be essentially only a rearrangement of the coordinates of P, with pairs of these latter interchanged and suitable arrangements as to signs.

$G(k, m)$ is a variety whose dimension d and order ν are given by

$$d = (k+1)(m-k), \quad \nu = \frac{1!\,2!\ldots k!}{(m-k)!\,(m-k+1)!\ldots m!}\, d!.$$

As examples we may mention

(*a*) the well-known representation of the lines of S_3 by the points of a (Klein) quadric W_4^2 of S_5,

(*b*) the mapping of the planes of S_5 on the points of a variety V_9^{42} of S_{19}.

As regards Grassmannians $G(k, m)$ in general, we recall the following facts:

(i) Every Grassmannian $G(k, m)$, by a theorem due to Severi [**23**], is such that the quadrics containing it have no other common point.

(ii) A pencil of S_k's of S_m (the S_k's that lie in a $[k+1]$ and pass through a $[k-1]$) is represented on $G(k, m)$ by a line; and, more generally:

(iii) There exist on $G(k, m)$ two families of (maximal) linear spaces, namely (*a*) spaces $[k+1]$ which each represent all the S_k's of S_m that lie in a fixed $[k+1]$, and (*b*) spaces $[m-k]$ which each represent the S_k's of S_m that pass through a fixed $[k-1]$.

As regards the type of Grassmannian with which we shall be mainly concerned, namely the model $G(n-1, 2n-1)$ of all the (self-dual) spaces $[n-1]$ of S_{2n-1}, we note:

(iv) A Grassmannian $G(n-1, 2n-1)$, being of dimension n^2, is multiply generated by two families of $[n]$'s, each of dimension n^2-1, such that any $[n]$ of the first family represents the $[n-1]$'s of S_{2n-1} that lie in a fixed $[n]$, while any $[n]$ of the second family represents the $[n-1]$'s of S_{2n-1} that pass through a fixed $[n-2]$.

6. Quadratic Veroneseans

Another kind of variety to which we shall frequently make reference in our investigations is that which we call the *quadratic Veronesean transform* of a given variety V, this being simply the projective model of all sections of V by quadrics of the ambient space. More precisely:

If $\xi = (\xi_i)$ $(i = 0, \ldots, m)$ is the generic point of an irreducible algebraic variety V in S_m, and if $M = \frac{1}{2}(m+1)(m+2)-1$, then the variety V' of S_M with generic point $\eta = (\xi_i \xi_j)$ $(i, j = 0, \ldots, m)$ is called the *quadratic Veronesean* of V.

The points of V' are in unexceptional birational correspondence with those of V; and if V is non-singular, so is V'. Further, every subvariety V_1' of V' is the quadratic Veronesean of the corresponding subvariety V_1 of V. If V is of dimension d and order ν, then V' (also of dimension d) is of order $\nu' = 2^d \nu$.

If V is a linear space, say S_m itself, then V' is the projective model of the complete system of quadrics of S_m – often referred to as the *Veronesean of quadrics* of S_m, or simply as the *quadric Veronesean* of S_m. It is evidently a variety V'_m of order 2^m whose ambient space is S_M; and the systems of linear subspaces of S_m are transformed into systems of sub-Veroneseans of V'_m. For $m = 1$, 2 and 3, the quadric Veroneseans of S_m are a conic, a Veronese surface ${}^0F_2^4[5]$ and a quadric Veronesean threefold $V_3'^8[9]$ respectively; and we note, for example, that $V_3'^8[9]$ contains an ∞^3-system (of grade 1) of Veronese surfaces corresponding to the planes of S_3.

The other kind of Veronesean variety with which we shall be concerned is the quadratic Veronesean of a non-singular quadric V_{2n-2}^2 of S_{2n-1}, projective model of all sections of V_{2n-2}^2 by all the other quadrics of S_{2n-1}. This is a variety V'_{2n-2} of order 2^{2n-1} whose ambient space is an S_{M-1}, where $M = n(2n+1)-1$. It contains two systems of quadric Veronesean $(n-1)$-folds, each of dimension $\frac{1}{2}n(n-1)$, which correspond to the two systems of generating $[n-1]$'s on V_{2n-2}^2. Thus, for example, the quadratic Veronesean of a quadric surface V_2^2 of S_3 is an octavic del Pezzo surface ${}^1F_2^8$ of the second kind in S_8; and ${}^1F_2^8$ contains two pencils of conics corresponding to the two systems of generators on V_2^2. Or, again, the quadratic Veronesean of a quadric V_4^2 of S_5 is a 4-fold V'_4 of order 32 in S_{19}; and this contains two ∞^3-systems of Veronese surfaces corresponding to the two systems of generating planes of V_4^2.

Chapter 3

CLIFFORD PARALLEL SPACES AND CLIFFORD REGULI

From now onwards, we shall be concerned always with geometry of S_{2n-1} relative to an 'absolute' non-singular quadric primal Ω; and in many of our statements therefore, to avoid tiresome repetition, we shall leave the qualifying phrase 'relative to Ω' to be understood by the reader. In the present Chapter we introduce the general notion of a pair of $[n-1]$'s of S_{2n-1} that are Clifford parallel relative to Ω, and we then discuss in detail the notion of a Clifford regulus as the simplest kind of 1-dimensional system of mutually Clifford parallel $[n-1]$'s of S_{2n-1}.

1. Clifford parallel $[n-1]$'s in S_{2n-1}

Since the definition of Clifford parallel spaces that we propose to adopt will strictly apply only to $[n-1]$'s which meet Ω in non-singular quadrics, we introduce for convenience the following

DEFINITION. An $[n-1]$ of S_{2n-1} will be termed *chordal* (relative to Ω) if it meets Ω in a non-singular quadric (a pair of distinct points if $n = 2$).

Equivalently, we may say that a chordal $[n-1]$ is one which does not touch Ω; or, again, one which does not meet its polar $[n-1]$.

DEFINITION. If Π is a chordal $[n-1]$, then a line which meets Π will be said to be *perpendicular to* Π (relative to Ω) if it also meets the polar space Π' of Π.

Such a line will then also be perpendicular to Π', and all the lines perpendicular to Π will evidently form a linear congruence since a unique transversal to Π and Π' can be drawn through a general point of S_{2n-1}.

Suppose now that Π_1 and Π_2 are a pair of skew chordal $[n-1]$'s and that the polar space Π_2' of Π_2 is also skew to Π_1. Then the perpendiculars to Π_2 from points of Π_1 are the transversal lines of Π_1, Π_2 and Π_2'. By Ch. 2, Prop. 3.2, however, these transversals are the directrix lines of a regulus R with Π_1, Π_2, Π_2' as generators. If it now happens that the polar space Π_1' of Π_1 is also a generator of R, then the directrix lines of R are also perpendicular to Π_1, i.e. Π_1 and Π_2 admit an ∞^{n-1} system of common perpendicular transversals, one through each point of either space. This motivates the

DEFINITION.† Two $(n-1)$-dimensional spaces Π_1 and Π_2 of S_{2n-1} will be said to be *Clifford parallel* (*relative to* Ω) if and only if they and their polar spaces Π_1' and Π_2' are four distinct generators of a regulus R of $[n-1]$'s of S_{2n-1}. They then admit ∞^{n-1} common perpendicular transversals – the directrices of R.

This definition implies, as will be noted, that Π_1, Π_2, Π_1', Π_2' are all chordal and mutually skew; also that if Π_1 is Clifford parallel to Π_2 then it is also Clifford parallel to Π_2'. A space Π cannot be Clifford parallel either to itself‡ or to its polar space Π'; but for convenience in stating certain results we add the subsidiary

DEFINITION. A chordal $[n-1]$ will be said to be *trivially parallel* both to itself and to its polar $[n-1]$.

2. Clifford reguli

We now propose to characterize analytically the aggregate of $[n-1]$'s of S_{2n-1} that are Clifford parallel to a given chordal space Π_0. To this end we choose a coordinate system, as by Ch. 2, §2 we may, so that the equation of Ω is $x^T x + y^T y = 0$ and the reference spaces X and Y are Π_0 and its polar space Π_0'. In these circumstances, since any $[n-1]$ which is Clifford parallel

† This definition is an almost word-for-word generalization of Clifford's original definition of parallelism (cf. Ch. 1, §3).

‡ The relation of Clifford parallelism is not only non-reflexive but also, as we shall see shortly, essentially non-transitive.

to X must be skew both to X and Y, it must have an equation of the form $y = Ax$ with $|A| \neq 0$; and the required condition on the matrix A is given by

PROPOSITION 2.1. *When the equation of* Ω *is* $x^T x + y^T y = 0$, *the necessary and sufficient condition that the* $[n-1]$ *given by* $y = Ax$ *should be Clifford parallel to the reference space* X *is*

$$A^T A = kI \quad (k \neq 0, -1), \tag{2.1}$$

where k *is a scalar.*

Proof. Let Π be the $[n-1]$ with equation $y = Ax$. If Π is Clifford parallel to X then (i) Π is skew to X, so that $|A| \neq 0$, (ii) Π is chordal, so that $|A^T A + I| \neq 0$ (cf. Ch. 2, §2(i)), and (iii) the polar space Π' of Π must belong to the regulus R containing X, Y and Π.

By Ch. 2, §2(ii) the equation of Π' is $y = -A^{-1T}x$; and by Ch. 2, §3 the generators of R are given, for different values of the scalar parameter λ, by $y = \lambda Ax$. Hence Π' is a generator of R if and only if

$$-A^{-1T} = \lambda A \tag{2.2}$$

for some value of λ ($\neq \infty$); and the values 0 and 1 of λ are excluded by (i) and (ii) above. A rearrangement of (2.2) now gives (2.1). Also, conversely, if A satisfies (2.1) for some $k \neq 0, -1$, then both A and $A^T A + I$ are non-singular so that Π is skew to X and chordal, and the previous argument then shows that Π is Clifford parallel to X. This completes the proof.

Now consider, in the notation of the above proof, the properties of the regulus R containing X, Y, Π and Π', and represented by the equation

$$y = \lambda Ax. \tag{2.3}$$

If Π_λ is the general $[n-1]$ of R, then its polar space $(\Pi_\lambda)'$ has equation $y = -(\lambda A^T)^{-1}x$, and this can be written by use of (2.1) in the form $y = \lambda' Ax$, where $\lambda' = -(k\lambda)^{-1}$. Thus $(\Pi_\lambda)'$ is the generator $\Pi_{\lambda'}$ of R, where λ, λ' are connected by the involutory relation

$$\lambda\lambda' = -1/k. \tag{2.4}$$

This means, then, that R is *autopolar* for Ω, in the sense that the polar of every generator of R is also a generator of R; and this further implies that every pair of chordal generators of R are either Clifford parallel or (if they are a polar pair) trivially parallel. Finally the two generators of R with parameter values given by $\lambda^2 = -1/k$ are self-polar for Ω, i.e. they lie on Ω; and these two generators separate harmonically in R all the polar pairs of generators of the regulus. These considerations lead to the

DEFINITION. A regulus of $[n-1]$'s of S_{2n-1} which is autopolar for Ω but is not entirely contained in Ω will be called a *Clifford regulus*. The two (distinct) generators of such a regulus that lie on Ω will be called its *terminal generators*.

We make use of the above terminology to recapitulate our results in

THEOREM 2.2. *If two $[n-1]$'s of S_{2n-1} are Clifford parallel, there exists a unique Clifford regulus R which contains them both. The two terminal generators of such a regulus (necessarily distinct and skew) lie on Ω and separate harmonically (within R) all the mutually polar pairs of generators of R. Apart from these two terminal generators, all the other generators of R are chordal and they are all either Clifford parallel or trivially parallel to each other.*

If two spaces Π_1 and Π_2 are Clifford parallel we shall often refer to the Clifford regulus containing them as that which *joins* Π_1 to Π_2. As regards the terminal generators of a Clifford regulus R, we note that these – being necessarily skew – will be generators of the same system or of opposite systems on Ω according as n is even or odd (Ch. 2, § 2, (*a*), (*b*), (*c*)). Thus when n is even Clifford reguli will be of two kinds – *α-reguli* and *β-reguli* – while if n is odd there will be no such distinction. Again, for n even, a pair of Clifford parallels may be called *α-parallel* or *β-parallel* according as the Clifford regulus to which they belong is an α-regulus or a β-regulus. For n odd no such distinction arises.

EX. 1. In the case of real elliptic geometry, for which S_{2n-1} is a real projective space and Ω is a virtual quadric, interpret and justify the statements: (i) a pair of Clifford parallel $[n-1]$'s are everywhere equidistant, and (ii) distance as measured between the generators of a Clifford regulus is additive. Property (ii) explains why Wong [**27**] refers to Clifford reguli as 'additive sets'.

Ex. 2. For the case $n = 2$, show that two lines of S_3 are Clifford parallel if they both meet the same pair of generators of one of the two systems on Ω. The two lines are α-parallel if they meet the same pair of β-generators; and *vice-versa.*

Show also in this case that a Clifford regulus is one regulus of a quadric Q which (a) meets Ω in a skew quadrilateral and (b) is such that Q and Ω are mutually apolar.

If Q satisfies the condition (a) but not (b), show that either regulus on Q is *autoparallel* but not autopolar, i.e. that any two members of it, neither a generator of Ω, are Clifford parallel. [N.B. This illustrates the general point that the autopolar property is, but the autoparallel property is not characteristic of a Clifford regulus.]

Ex. 3. With the notation of the present section, show that every $[n-1]$ Clifford parallel to X is a generator of some quadric of the pencil $x^T x + \lambda y^T y = 0$; also, conversely, that any $[n-1]$-generator of such a quadric that is both chordal and skew to X and Y is Clifford parallel to X. Deduce that the $[n-1]$'s Clifford parallel to a given $[n-1]$ have freedom $\frac{1}{2}(n^2 - n + 2)$.

Show further that the $[n-1]$'s parallel to a given $[n-1]$ that pass through a general point P of S_{2n-1} belong to one or other of two irreducible systems, each of freedom $\frac{1}{2}(n-1)(n-2)$. Discuss the case $n = 2$.

3. Questions of transitivity

The generally non-transitive character of Clifford parallelism follows from

PROPOSITION 3.1. *If, in the notation of the preceding section,* $y = A_1 x$ *and* $y = A_2 x$ *are the equations of two spaces* Π_1 *and* Π_2 *which are each Clifford parallel to the reference space* X, *so that*

$$A_1^T A_1 = k_1 I \quad \text{and} \quad A_2^T A_2 = k_2 I \quad (k_1, k_2 \neq 0, -1), \tag{3.1}$$

then a necessary and sufficient further *condition for* Π_1 *to be Clifford parallel to* Π_2 *is*

$$A_1^T A_2 + A_2^T A_1 = kI \quad (k \neq k_1 + k_2 \text{ or } -(1 + k_1 k_2)). \tag{3.2}$$

Proof. Suppose first that Π_1 is Clifford parallel to Π_2. Then the polars Π'_1 and Π'_2 of these spaces have equations $y = -A_1^{-1T}x$ and $y = -A_2^{-1T}x$; and since Π_1, Π_2, Π'_1, Π'_2 belong to a regulus, we find, by a straightforward application of the matrix cross-ratio theorem (Ch. 2, Th. 4.1) and use of the relations (3.1), that A_1 and A_2 satisfy (3.2).

Now suppose conversely that A_1 and A_2 satisfy (3.2) in addition to (3.1). For the reversion of the algebra, applying the sufficiency part of Th. 4.1 of Ch. 2, it is only necessary to verify that (3.1) and (3.2) together imply that the four spaces Π_1, Π_2, Π'_1, Π'_2 are all mutually skew. By (3.1) we know that Π_1 and Π'_1 are skew, as are also Π_2 and Π'_2.

To verify that Π_1 and Π_2 are skew – and therefore also Π'_1 and Π'_2 – we observe that

$$\begin{aligned}(A_1 - A_2)^T (A_1 - A_2) &= A_1^T A_1 + A_2^T A_2 - (A_1^T A_2 + A_2^T A_1)\\ &= (k_1 + k_2 - k) I,\end{aligned}$$

from which it follows, since $k \neq k_1 + k_2$, that $A_1 - A_2$ is non-singular and hence Π_1 and Π_2 are skew.

Similarly, to see that Π_1 and Π'_2 are skew – and hence also Π_2 and Π'_1 – we observe that

$$\begin{aligned}(A_1 + A_2^{-1T})^T (A_1 + A_2^{-1T}) &= (A_1^T + k_2^{-1} A_2^T)(A_1 + k_2^{-1} A_2)\\ &= A_1^T A_1 + k_2^{-2} A_2^T A_2\\ &\qquad + k_2^{-1}(A_1^T A_2 + A_2^T A_1)\\ &= k_2^{-1}(k_1 k_2 + 1 + k) I,\end{aligned}$$

from which, since $k \neq -(1 + k_1 k_2)$, it follows that $A_1 + A_2^{-1T}$ is non-singular and hence Π_1 and Π'_2 are skew.

It follows then, since Π_1, Π_2, Π'_1, Π'_2 are all mutually skew, that the sufficiency section of Th. 4.1 of Ch. 2 can be applied to establish the sufficiency of the conditions (3.1) and (3.2) and this completes the proof.

From the above we derive the following theorem which will be important in the sequel.

THEOREM 3.2. *If* Π, Π_1 *and* Π_2 *are any three mutually Clifford parallel* $[n-1]$*'s of* S_{2n-1}, *then* Π_2 *is Clifford parallel to almost all generators of the Clifford regulus joining* Π *to* Π_1.

Proof. Retaining the previous notation, we take Π to be the reference space X and Π_1, Π_2 to have equations $y = A_1 x$ and $y = A_2 x$ respectively. Then A_1 and A_2, by hypothesis, satisfy conditions (3.1) and (3.2) of Prop. 3.1.

Now let R be the Clifford regulus joining X to Π_1, and let Π_λ be a generator $y = \lambda A_1 x$ of R which is Clifford parallel to X. This excludes X and Y (which are already Clifford parallel to Π_2) and the two terminal generators of R (given by $\lambda^2 = -1/k_1$). By Prop. 3.1 the conditions for Π_λ to be Clifford parallel to Π_2 are those obtained by replacing A_1 by λA_1 in (3.1) and (3.2) with the consequent replacement of k_1 by $\lambda^2 k_1$ and of k by λk. It follows then that Π_λ is Clifford parallel to Π_2 provided only that

(i) $\lambda k \neq \lambda^2 k_1 + k_2$, and

(ii) $\lambda k \neq -(1+\lambda^2 k_1 k_2)$,

the condition $\lambda^2 \neq -1/k_1$ having already been imposed. There are, therefore, at most six generators of R which are not Clifford parallel to Π_2, namely the two terminal generators of R, the two generators given by (i), which are those not skew to Π_2, and the two generators given by (ii), which are those not skew to the polar space Π_2' of Π_2. This proves the theorem.

It appears incidentally from our analysis that Π_2 meets at most two generators of the regulus joining Π to Π_1, and moreover it meets them in linear spaces of dimension† at least $\frac{1}{2}n - 1$.

COROLLARY 3.2.1. *If $\Pi_1, \ldots, \Pi_4$ are four mutually Clifford parallel $[n-1]$'s of S_{2n-1}, then the general space of the Clifford regulus joining Π_1 to Π_2 is Clifford parallel to the general space of the Clifford regulus joining Π_3 to Π_4.*

For if Π is the general space of the first regulus, then the Theorem tells us that Π is Clifford parallel to Π_3 and Π_4, and then, further, that Π is Clifford parallel to the general space of the second regulus.

† Since $(\lambda A_1 - A_2)^T (\lambda A_1 - A_2) = (\lambda^2 k_1 - \lambda k + k_2) I$, it follows that $\lambda A_1 - A_2$ is singular if and only if $\lambda^2 k_1 - \lambda k + k_2 = 0$; and this is the condition that Π_λ meets Π_2. Moreover, by a well-known inequality concerning the rank of a product of two matrices, it follows that if $\lambda A_1 - A_2$ is singular then its rank is at most $\frac{1}{2}n$; and this means that Π_2 meets Π_λ in a space of dimension at least $\frac{1}{2}n - 1$.

4. Canonical equation of a Clifford regulus

We now propose to find all the Clifford reguli which have a given pair of skew $[n-1]$-generators of Ω as their terminal spaces.

For this purpose we use the second standard form for the equation of Ω, choosing the coordinate system, as by Ch. 2, § 2 we may, so that the two given $[n-1]$'s are the reference spaces X and Y, and the equation of Ω is $x^T y = 0$. If Π is the $[n-1]$ with equation $y = Ax$ (so that Π does not meet Y), we note first that

(i) the polar space Π' of Π has equation $y = -A^T x$, and

(ii) Π is chordal if and only if $A + A^T$ is non-singular.

We now prove

PROPOSITION 4.1. *If A is any symmetric non-singular $n \times n$ matrix, then the equation*

$$y = \lambda A x \tag{4.1}$$

defines (in the usual way) a Clifford regulus with X and Y as terminal generators. Conversely, every such Clifford regulus is defined by an equation (4.1) in which A is symmetric and non-singular.

Proof. If A is symmetric and non-singular, then $A + A^T = 2A$ is non-singular; whence, by (ii) above, (4.1) defines a regulus R of which every member except X and Y is chordal. Further, since $A = A^T$, the generators of R with parameters $\pm\lambda$ are always polars of one another (cf. (i) above). Hence R is a Clifford regulus.

Conversely, if R is a Clifford regulus with X and Y as terminal spaces, then it is defined by an equation $y = \lambda A x$, $|A| \neq 0$, and it is such that any two of its generators given by values $\pm\lambda$ of the parameter must be polars of each other. This shows that $A = A^T$, so that A is symmetric; and this completes the proof.

COROLLARY 4.1.1. *The freedom of Clifford reguli with assigned terminal generators is $\frac{1}{2}(n-1)(n+2)$, and that of all Clifford reguli is $\frac{1}{2}(n-1)(3n+2)$.*

We now establish an important canonical form to which the equations of Ω and any given Clifford regulus can be simultaneously reduced.

THEOREM 4.2. *Being given a Clifford regulus R in S_{2n-1}, we can always choose a coordinate system in which (simultaneously) Ω has equation $x^T y = 0$ and R is given by $y = \lambda x$.*

Proof. By taking the terminal generators of R as the reference spaces X and Y, we can certainly reduce the equation of Ω to the form $x^T y = 0$, and then, by Prop. 4.1, R is given by an equation $y = \lambda A x$, where A is symmetric and non-singular. Now there exists a non-singular matrix B such that $A = BB^T$ (since any non-singular symmetric matrix can be so written). In terms of new coordinates (ξ, η) defined by the allowable transformation

$$x = A^{-1}B\xi, \quad y = B\eta,$$

we find that the new equation of Ω is $\xi^T \eta = 0$ and that R is now given by the equation $\eta = \lambda\xi$. This proves the Theorem.

Ex. 1. If Π_0, Π_0' is any (chordal) polar pair of generators of a Clifford regulus R, then they define an involutory collineation $\varpi(\Pi_0, \Pi_0')$ of S_{2n-1} – a biaxial harmonic homology with Π_0 and Π_0' as spaces of united points – which leaves Ω invariant. Prove that $\varpi(\Pi_0, \Pi_0')$ permutes the generators of R among themselves, carrying polar pairs into polar pairs. (Analytically, if Ω and R are taken in the canonical forms given in Th. 4.2, and if Π_0 and Π_0' are given by the parameter values $\lambda = \pm\lambda_0$, then $\varpi(\Pi_0, \Pi_0')$ carries the generator with parameter λ into that with parameter λ_0^2/λ.)

Show also that, if Π_1 and Π_2 are any two distinct chordal generators of R, then there exists a unique polar pair Π_0, Π_0' of generators of R for which the associated homology $\varpi(\Pi_0, \Pi_0')$ carries Π_1 into Π_2.

5. The kernel of a Clifford regulus

Let R be a Clifford regulus and let X, Y denote its terminal generators. Since R, as an algebraic variety, is a V_n^n, it meets Ω (by the Theorem of Bézout) in an $(n-1)$-fold of total order $2n$; but part of this intersection consists of the two terminal generators X, Y, and the remainder is therefore a V_{n-1}^{2n-2}. We propose to discuss the character of this latter variety.

We observe first that any chordal generator Π of R meets Ω in an $(n-2)$-dimensional quadric q; and the directrix line of R through any point of q meets Ω in at least two further points, namely the intersections of the line with X and Y. Thus this line lies entirely on Ω and meets the section of Ω by every chordal generator of R. It follows then that ∞^{n-2} of the directrix lines of R lie on Ω, that these generate the V_{n-1}^{2n-2} referred to above and that they meet every chordal generator of R in the section of Ω by that generator. They will also meet X and Y in $(n-2)$-dimensional quadrics q_X and q_Y which we may call the *terminal quadrics of* R. Recapitulating we have

PROPOSITION 5.1. *If R is a Clifford regulus of S_{2n-1}, then the ∞^{n-2} directrix lines of R that lie on Ω generate a variety V_{n-1}^{2n-2}, the intersection of R with Ω residual to the two terminal generators X, Y of R. These directrix lines meet every chordal generator of R in the section q of that generator by Ω, and they meet X and Y in the terminal quadrics q_X and q_Y of R. All the quadrics q, including q_X and q_Y, lie over any one of themselves in R.*

This leads to the

DEFINITION. *The kernel of a Clifford regulus R* of S_{2n-1} is the V_{n-1}^{2n-2} generated by the ∞^{n-2} directrices of R that lie on Ω, and it will be denoted by the symbol K_R. These same lines will be called the *directrix lines* of K_R, and the non-singular quadrics q in which they meet the generators of R will be called the *quadric-generators* of K_R.

The existence of the terminal quadrics q_X and q_Y of R leads to a further definition which will be of considerable importance in the sequel.

DEFINITION. A couple (Π, q), where Π is an $[n-1]$-generator of Ω and q is a non-singular $(n-2)$-dimensional quadric in Π, will be called an *augmented generator of* Ω. We shall find it more convenient to denote the couple in question by the symbol $\Pi(q)$.

An augmented generator, as will appear, plays the part of a pair of mutually polar $[n-1]$'s that have tended to coincidence in a particular way (within a Clifford regulus).

To obtain equations defining K_R, we may suppose, by Th. 4.2, that Ω and R are given in the canonical forms $x^T y = 0$ and $y = \lambda x$. Then K_R is the locus of points (x, y) that satisfy for some value of λ the equations

$$x^T x = y^T y = 0, \quad y = \lambda x, \tag{5.1}$$

and the terminal quadrics q_X and q_Y (corresponding to $\lambda = 0$ and $\lambda = \infty$) are given by

$$q_X\colon \quad x^T x = 0, \quad y = 0,$$

$$q_Y\colon \quad y^T y = 0, \quad x = 0.$$

Now equations (5.1) are readily identified as those of a Segre product of a non-singular $(n-2)$-dimensional quadric with a line – a variety known to be a V_{n-1}^{2n-2} of S_{2n-1} (irreducible for $n > 2$, a pair of skew lines when $n = 2$). This gives

PROPOSITION 5.2. *The kernel K_R of a Clifford regulus R is a Segre product of a non-singular $(n-2)$-dimensional quadric with a line. The terminal generators X and Y of R take on, by the adjunction of the terminal quadrics q_X and q_Y respectively, the character of augmented generators of Ω.*

We now note briefly some further – and easily verifiable – properties of a Clifford regulus R and its kernel K_R.

(i) If d is any directrix line of K_R, and if P is the point where d meets any quadric generator q of K_R, then the polar secundum of d (with respect to Ω) meets the generator of R containing q in the tangent $[n-2]$ to q at P.

(ii) In the notation of (i), as P moves along d the tangent $[n-2]$ to q at P describes a regulus of $[n-2]$'s in the polar secundum of d.

(iii) If q_1, q_2 are any two quadric generators of K_R which lie in generators Π_1, Π_2 of R that are not mutually polar, then the tangent primes to Ω at points of q_1 all touch q_2, i.e. q_2 is the envelope of the $[n-2]$'s in which tangent primes to Ω at points of q_1 meet Π_2.

(iv) In particular, the terminal quadrics of R lie over each other in R in such a way that the tangent prime to Ω at any point

P of either of them touches the other at the point of the latter which lies over P.

This last statement motivates the further

DEFINITION. Two augmented generators $\Pi_1(q_1)$ and $\Pi_2(q_2)$ of Ω will be said to be *compatible* if there exists a Clifford regulus with Π_1, Π_2 as its terminal generators and q_1, q_2 as its terminal quadrics.

In amplification of (iv) then, we leave it to the reader to verify

PROPOSITION 5.3. *Two augmented generators* $\Pi_1(q_1)$ *and* $\Pi_2(q_2)$ *of* Ω *are compatible if and only if* (i) Π_1 *is skew to* Π_2, *and* (ii) *the (non-singular) quadrics* q_1, q_2 *are such that* q_2 *is the envelope of the sections of* Π_2 *by the tangent primes to* Ω *at the points of* q_1.

Ex. 1. Prove that two $[n-1]$'s Π_1 and Π_2 of S_{2n-1} are Clifford parallel if and only if (i) they and their polar spaces are all skew to one another, and (ii) they meet Ω in quadrics q_1 and q_2 such that the tangent primes to Ω at points of q_1 all touch q_2. (These conditions then imply that the tangent primes to Ω at points of q_2 all touch q_1.)

Ex. 2. Show that a Clifford regulus is uniquely determined by its terminal generators and one of its terminal quadrics; and hence confirm again (cf. Cor. 4.1.1) that the Clifford reguli with given terminal generators have freedom $\frac{1}{2}(n-1)(n+2)$.

Ex. 3. If Ω has the equation $x^T y = 0$, then any regulus R with the reference spaces X, Y as two of its generators is given by an equation $y = \lambda A x$ with $|A| \neq 0$; and if R is not entirely contained in Ω it meets Ω residually in a V_{n-1}^{2n-2} which is a Segre product of an $(n-2)$-dimensional quadric with a line. If d is any directrix line of V_{n-1}^{2n-2} and P is the point in which d meets a variable quadric generator q of V_{n-1}^{2n-2}, show that the tangent $[n-2]$ to q at P describes a regulus R_{n-1}^{n-1} lying in a certain secundum through d, but that this is not the polar secundum of d with respect to Ω unless $A = A^T$ so that R is a Clifford regulus.

Show further that if R is not autopolar $(A \neq A^T)$ then it is at least autoparallel if A satisfies the condition

$$A^T A^{-1} + A A^{-1T} = kI$$

for some value of k, a condition which holds always if $n = 2$ but not if $n > 2$.

Ex. 4. If κ is the natural correlation between two skew generators X, Y of Ω (cf. Ch. 2, § 2) and ϖ is a non-singular collineation between X and Y, show that the regulus generated by ϖ is a Clifford regulus R (with X and Y as terminal generators) if and only if the correspondence $\kappa^{-1}\varpi$ in X is a symmetrical polarity. (It will then, of course, be the correspondence of pole and polar with respect to the terminal quadric in X of the Clifford regulus R.) The same condition may be expressed by saying that ϖ must be of the form $\kappa\tau$, where τ is a non-singular symmetric polarity in X.

Ex. 5. If R is the regulus $y = \lambda x$ and A is any non-singular matrix of order n, show that R is a Clifford regulus with respect to the quadric $x^T A y = 0$ if and only if A is symmetric.

Show also that, if C_0 is a fixed non-singular symmetric matrix and S is an arbitrary skew-symmetric matrix, then all the quadrics $x^T(C_0 + S)y = 0$ meet R in the same kernel variety V_{n-1}^{2n-2} (residual to X and Y), but R is a Clifford regulus with respect to only one of them, namely $x^T C_0 y = 0$.

Ex. 6. If n is even and S_1, S_2 are skew-symmetric matrices such that $S_1 - S_2$ is non-singular, show that

$$(y - S_1x)^T(S_1 - S_2)^{-1}(y - S_2x) \equiv x^T y$$

and hence, using Prop. 4.1, write down equations for a generic pair of α-parallels.

Similarly, if n is odd and S, T are skew-symmetric matrices such that $I - ST$ is non-singular, show that

$$(x - Ty)^T(I - ST)^{-1}(y - Sx) \equiv x^T y$$

and hence write down equations for a generic pair of Clifford parallels (n odd).

6. Ω-reguli

Besides the Clifford reguli so far considered, we shall also on occasion be concerned with reguli whose generators all lie on

Ω – to be called *Ω-reguli*. Since the generating $[n-1]$'s of either system on Ω are not skew to each other if n is odd, we note that Ω-reguli can only exist in S_{2n-1} when n is even. Further, when n is even, Ω-reguli will be of two kinds, consisting respectively of α-generators and β-generators (of the two systems of generating $[n-1]$'s of Ω). We now assert

PROPOSITION 6.1. *If S is any skew-symmetric non-singular $n \times n$ matrix (so that n is necessarily even) and if the equation of Ω is $x^T y = 0$, then the equation*

$$y = \lambda S x \tag{6.1}$$

defines an Ω-regulus with the reference spaces X and Y as two of its generators. Conversely every such Ω-regulus is defined by an equation (6.1) *in which S is skew-symmetric and non-singular.*

Proof. The proof follows at once from the observations that the space with equation $y = Ax$ is skew to X and Y if and only if $|A| \neq 0$, and it lies on Ω if and only if A is skew-symmetric (cf. Ch. 2, §2).

From the above it follows that (i) the Ω-reguli which contain X and Y have freedom $\frac{1}{2}(n-2)(n+1)$, and (ii) the total freedom of Ω-reguli of either system is $\frac{3}{2}(n-2)(n+1)$.

Ex. 1. With the notation of Ex. 4 of the preceding section, show that the regulus generated by a non-singular collineation ϖ between X and Y is an Ω-regulus if and only if the correlation $\kappa^{-1}\varpi$ in X is a (non-singular) skew-polarity. In other words, ϖ must be of the form $\kappa\sigma$, where σ is a non-singular skew polarity in X.

In this case show that a directrix line of the Ω-regulus and its polar secundum meet X in a point and an $[n-2]$ which correspond by σ.

CHAPTER 4

LINEAR SYSTEMS OF CLIFFORD PARALLELS

The general topic of this chapter is the structure and properties of irreducible systems of $[n-1]$'s of S_{2n-1}, of any dimension $r \geqslant 1$, which have the property that their members are in general† Clifford parallel to one another with respect to the absolute quadric Ω.

Among such systems we introduce a special class which we call linear, with the property that every irreducible system of parallels is embedded in a linear system. We then reduce the construction of linear systems of parallels to that of finding solutions to a set of matrix equations—an extension of the Hurwitz–Radon matrix equations. This gives us a canonical form for the equation of any linear system which we study in some detail, especially as regards certain properties which lead us to distinguish a particular kind of linear system, to be called non-singular, as being of maximum interest; and we show how to classify these non-singular linear systems. In a final section we discuss the question as to whether linear systems of parallels with respect to one absolute quadric can also be linear systems of parallels with respect to other absolute quadrics.

1. Clifford parallel and algebraically parallel spaces

Let $\mathscr{G}$ be the irreducible algebraic system of all $[n-1]$'s of S_{2n-1}; and let $\mathscr{G} * \mathscr{G}$ denote the symmetric product of $\mathscr{G}$ by itself, i.e. the irreducible algebraic system of all the *unordered* pairs of spaces of $\mathscr{G}$. We now wish to consider irreducible subsystems of $\mathscr{G} * \mathscr{G}$ such that the generic pair of any one of these subsystems is a pair of Clifford parallel $[n-1]$'s in the strict sense in which we defined this concept in Ch. 3.

† In a system of $[n-1]$'s of dimension $r > 1$, each space will be met by ∞^{r-1} others, so that we cannot expect to find systems of dimension greater than 1 in which every pair of chordal spaces are Clifford (or trivially) parallel. The best that can be hoped for are systems in which *almost all* pairs of spaces are Clifford parallel.

In this connection, however, we must first point out that Clifford parallel pairs in S_{2n-1} are all of one kind when n is odd but of two different kinds if n is even. For, by Th. 2.2 of Ch. 3, the $[n-1]$'s of any Clifford parallel pair are joined by a unique Clifford regulus R; and whereas, if n is odd, R is of the unique type with terminal generators of opposite systems α and β of Ω, we have already noted (Ch. 3, § 2) that, if n is even, R can be either an α-regulus (with terminals α_1, α_2) or a β-regulus (with terminals β_1, β_2). This means then that whereas, for n odd, we can regard all pairs of Clifford parallels as specializations of a single generic pair (Π_1, Π_2), we have to regard all such pairs, when n is even, as specializations of one or other of two generic pairs – a generic pair $(\Pi_1, \Pi_2)^\alpha$ of α-parallels and a generic pair $(\Pi_1, \Pi_2)^\beta$ of β-parallels (cf. Ch. 3, § 5, Ex. 6).

Now suppose, for n odd, that $\mathscr{V}$ is the unique irreducible subsystem of $\mathscr{G} * \mathscr{G}$ defined by the generic parallel pair (Π_1, Π_2). We shall say in this case that two spaces Π_1^* and Π_2^* are *algebraically parallel* if (Π_1^*, Π_2^*) is *any* pair of $\mathscr{V}$, i.e. any specialization of (Π_1, Π_2). Similarly, for n even, if $\mathscr{V}_\alpha$ and $\mathscr{V}_\beta$ are the irreducible subsystems of $\mathscr{G} * \mathscr{G}$ with generic pairs $(\Pi_1, \Pi_2)^\alpha$ and $(\Pi_1, \Pi_2)^\beta$ respectively, we shall say that the spaces of any pair of $\mathscr{V}_\alpha$ are *algebraically α-parallel*, while those of any pair belonging to $\mathscr{V}_\beta$ are *algebraically β-parallel.*

Ex. 1. (*The case $n = 2$.*) The situation for n even can be illustrated for $n = 2$ as follows. If we map the lines of S_3 on the points of a Klein quadric W_4^2 of S_5, then the α-generators and β-generators of Ω are mapped on W_4^2 by the points of two conics k_1 and k_2 respectively – sections of W_4^2 by a skew pair of polar planes π_1 and π_2. We then find that

(i) a pair of α-parallel (resp. β-parallel) lines of S_3 (cf. Ch. 3, § 2, Ex. 2) is represented by a pair of distinct points P_1, P_2 of W_4^2, remote from k_1 and k_2, such that their join P_1P_2 does not lie on W_4^2 but meets π_1 (resp. π_2) and is skew to π_2 (resp. π_1),

(ii) the system $\mathscr{V}_\alpha$ (resp. $\mathscr{V}_\beta$) is represented on W_4^2 by the aggregate of point-pairs of W_4^2 whose joining lines (possibly tangent to W_4^2 or lying on it) meet π_1 (resp. π_2),

(iii) the two types of Clifford regulus in S_3 are represented on W_4^2 by the (proper) conics in which W_4^2 is met by planes which meet one of π_1, π_2 in a line (not touching k_1 or k_2) and the other in a point (not lying on k_1 or k_2).

2. Systems of Clifford parallels

We now introduce the

DEFINITION. An *irreducible system of Clifford parallels* in S_{2n-1} is any irreducible system of $[n-1]$'s, defined over the complex ground field, which is such that a generic pair of its members is a Clifford parallel pair.

If $\mathscr{F}$ is such a system, the above definition implies that its symmetric product $\mathscr{F} * \mathscr{F}$ is contained in the unique irreducible system $\mathscr{V}$ if n is odd, or in one or other of $\mathscr{V}_\alpha$, $\mathscr{V}_\beta$ if n is even; and, further, that the spaces of a generic pair of $\mathscr{F}$ are strictly (not only algebraically) parallel. This again implies that almost all members of $\mathscr{F}$ are chordal; also that two members of $\mathscr{F}$ which do not meet each other or their polar spaces are Clifford parallel. A Clifford regulus is a special type of 1-dimensional irreducible system of Clifford parallels.

An irreducible system $\mathscr{F}$, as above defined, may or may not have the property that it contains the Clifford regulus joining every Clifford parallel pair of its members. This leads to the

DEFINITION. An irreducible system of Clifford parallels is said to be *linear* if it contains the Clifford regulus joining every Clifford parallel pair of its members.

We now prove

THEOREM 2.1. (*Embedding Theorem.*) *Every irreducible system of Clifford parallels is contained in at least one linear system of Clifford parallels.*

Proof. Let $\mathscr{F}$ be the given irreducible system which we suppose to be of dimension $r \geqslant 1$ and let Π_1 and Π_2 be two independent generic spaces of $\mathscr{F}$. Then Π_1 and Π_2, being Clifford parallel, belong to a Clifford regulus. Let Π denote a generic $[n-1]$ of this regulus, and let $\mathscr{F}'$ denote the irreducible system of $[n-1]$'s of S_{2n-1} with Π as generic member. Plainly, then, $\mathscr{F}'$ contains $\mathscr{F}$. Further, by Ch. 3, Cor. 3.2.1, any two independent generic members of $\mathscr{F}'$ are Clifford parallel. If $\mathscr{F}$ was linear to begin with, $\mathscr{F}'$ coincides with $\mathscr{F}$; but, otherwise, $\mathscr{F}'$ strictly contains $\mathscr{F}$, having dimension $r' > r$. Repeating the same argument we find that $\mathscr{F}'$, if it is not linear, is contained in an irreducible system $\mathscr{F}''$ (of Clifford parallels) of dimension $r'' > r'$. Continuation of this process must terminate since the sequence $r, r', r'', \ldots$ is bounded above by the dimension of the complete system of $[n-1]$'s of S_{2n-1}. Hence, after a finite number of steps, we must arrive at a linear system $\mathscr{F}^{(k)}$ which contains the original system $\mathscr{F}$. This proves the Theorem.

We add finally

PROPOSITION 2.2. *A linear system $\mathscr{F}$ of Clifford parallels is autopolar with respect to Ω, i.e. the polar of every space of $\mathscr{F}$ belongs to $\mathscr{F}$.*

The proof may be left to the reader.

3. The canonical equation of a linear system

To simplify the statement of the main theorem of this section we prefix the

DEFINITION. If $r \geqslant 2$, a set $S_1, \ldots, S_{r-1}$ of $r-1$ skew-symmetric matrices of order n will be said to satisfy the *extended Hurwitz–Radon matrix equations for given integers n, r and k* $(1 \leqslant k \leqslant r)$ if they are such that

$$\left.\begin{array}{lll}
\text{(i)} & S_i^T = -S_i & (i = 1, \ldots, r-1), \\
\text{(ii)} & S_i^2 = -I & \text{if} \quad i \leqslant k-1, \\
\text{(iii)} & S_i^2 = 0 & \text{if} \quad i > k-1, \\
\text{(iv)} & S_i S_j + S_j S_i = 0 & (i,j = 1, \ldots, r-1,\ i \neq j).
\end{array}\right\} \qquad (3.1)$$

The basic result we propose to establish may now be stated as

THEOREM 3.1. *If $\mathscr{F}$ is any r-dimensional linear system $(r \geqslant 1)$ of Clifford parallels in S_{2n-1}, then a coordinate system can be chosen* *in which the absolute quadric* Ω *has equation* $x^T y = 0$ *and almost all the* $[n-1]$*'s of* $\mathscr{F}$ *(excluding only those which meet the reference space* Y*) are given, for different sets of values of the parameters* $\alpha, \lambda_1, \ldots, \lambda_{r-1}$ *by an equation*

$$y = (\alpha I + \lambda_1 S_1 + \ldots + \lambda_{r-1} S_{r-1})\, x, \tag{3.2}$$

where $S_1, \ldots, S_{r-1}$ *is a set of* $r-1$ *linearly independent matrices of order* n *which satisfy the extended Hurwitz–Radon matrix equations* (3.1) *for some value of* k *in the range* $1 \leqslant k \leqslant r$.

Conversely, if $S_1, \ldots, S_{r-1}$ *is any such set of matrices, then* (3.2) *represents almost all the spaces of an r-dimensional linear system of Clifford parallels with respect to the quadric* Ω *given by* $x^T y = 0$.

For the purposes of the proof which follows, we note first that $\mathscr{F}$ must certainly contain pairs of Clifford parallel $[n-1]$'s whose coordinates belong to the ground field; and if we choose any one such pair (Π_1, Π_2), then $\mathscr{F}$ (being linear) must also contain the Clifford regulus R_0 joining Π_1 to Π_2. Further, by Ch. 3, Th. 4.2, we can then choose a coordinate system in which Ω has equation $x^T y = 0$ and R_0 has equation $y = \lambda x$. *In all that follows, therefore, we shall assume that*

(i) Ω *has equation* $x^T y = 0$, *and*

(ii) $\mathscr{F}$ *contains the Clifford regulus* R_0 *with equation* $y = \lambda x$.

The proof of the first part of Theorem 3.1 proceeds by a sequence of Lemmas involving successive applications of the matrix cross-ratio theorem (Ch. 2, Th. 4.1).

LEMMA A. *The space* $y = Ax$ *is Clifford parallel to almost all generators of* R_0 *if and only if* $A = \alpha I + T$, *where* α *is a non-zero scalar and* T *is a skew-symmetric matrix such that* $T^2 = tI$, *where* t *is a scalar.*

Proof. This Lemma is valid also when the elements of A belong to an extension of the (complex) ground field. Bearing this in mind, we suppose that $y = Ax$ is Clifford parallel to almost all generators of R_0; and then, if λ is an indeterminate over the field obtained by adjoining the elements of A to the ground field, the space $y = Ax$ will be Clifford parallel to $y = \lambda x$, and hence the four spaces

$$y = Ax, \quad y = -A^T x, \quad y = \lambda x, \quad y = -\lambda x$$

will belong to a regulus. They must therefore be mutually skew; and hence, by the matrix cross-ratio theorem, there must exist a scalar $c(\lambda)$, not equal to 0 or 1, such that†

$$\{\lambda I, A;\ -A^T, -\lambda I\} = c(\lambda) I.$$

On reduction, this asserts that A must satisfy the equation

$$\lambda^2 I + \lambda\{1 - 2c(\lambda)\}(A + A^T) + A^T A = 0 \qquad (3.3)$$

involving the indeterminate λ. Replacing λ by another indeterminate μ and subtracting, we see that $A + A^T$ is a scalar multiple of I; for the coefficient of $A + A^T$ after subtraction could only be zero if each of $\lambda\{1 - 2c(\lambda)\}$ and $\mu\{1 - 2c(\mu)\}$ had the same value k, independent of λ and μ, and then (3.3) could not be satisfied. Thus we may write $A + A^T = 2\alpha I$, i.e.

$$A = \alpha I + T,$$

where T is skew-symmetric and α is non-zero (as follows from equation (3.3) since λ is an indeterminate). It now follows from (3.3) that $A^T A$ is also a scalar multiple of I; whence, since

$$A^T A = (\alpha I - T)(\alpha I + T) = \alpha^2 I - T^2,$$

it follows that $T^2 = tI$, where t is a scalar.‡ Having thus proved the necessity of the condition given in the Lemma, we remark that its sufficiency now follows by direct computation; and the Lemma is therefore proved.

For completeness we note that the value of the scalar $c(\lambda)$ is $\{(\lambda + \alpha)^2 - t\}/(4\alpha\lambda)$. Further, we note that the condition of the Lemma does not exclude $T^2 = 0$ (i.e. $t = 0$).

LEMMA B. *If a linear system $\mathscr{F}$ contains both the regulus R_0 and the space $y = (\alpha I + T)x$ as defined in Lemma A, then it contains every space $y = (\delta I + \epsilon T)x$, where δ and ϵ are arbitrary scalars.*

† The order of the elements within the brackets on the left is easiest for computation.

‡ If $t = \alpha^2 \neq 0$, this implying that n is even and $A^T A = 0$, then the polar spaces $y = Ax$ and $y = -A^T x$ meet X in skew spaces $[\tfrac{1}{2}n - 1]$ (cf. §4, Ex. 2 below).

Proof. It is plainly enough to show that the *generic* space of the type in question belongs to $\mathscr{F}$; and, by the linearity of $\mathscr{F}$, this will follow if we show that the same space belongs to a Clifford regulus joining $y = (\alpha I + T)x$ to a generator of R_0. We therefore interpret δ and ϵ as independent indeterminates over the field obtained by adjoining α and the elements of T to the ground field, and we define μ (to within sign) by the equation

$$\delta^2 - \epsilon\alpha^2 = (1-\epsilon)(\mu^2 - \epsilon t).$$

By the matrix cross-ratio theorem, we find then that the four spaces

$$y = \mu x, \quad y = -\mu x, \quad y = (\alpha I + T)x, \quad \text{and} \quad y = (\delta I + \epsilon T)x$$

belong to a regulus R; and since the first and third of these are Clifford parallel, while the second is the polar of the first, it follows that R is a Clifford regulus. Hence $\mathscr{F}$, by its linearity, contains R and therefore, *a fortiori*, the space $y = (\delta I + \epsilon T)x$.

LEMMA C. *If a linear system $\mathscr{F}$ contains R_0 and the two spaces $y = (\alpha_1 I + T_1)x$ and $y = (\alpha_2 I + T_2)x$, these latter being (strictly) Clifford parallel, then there exists a scalar t_{12} such that*

$$T_1 T_2 + T_2 T_1 = 2t_{12} I.$$

Proof. We know, by Lemma A, that $\alpha_1 \neq 0$, $\alpha_2 \neq 0$ and T_1, T_2 are skew-symmetric and such that $T_1^2 = t_1 I$, $T_2^2 = t_2 I$. We express then the condition that the four spaces

$$y = (\pm\alpha_1 I + T_1)x, \quad y = (\pm\alpha_2 I + T_2)x$$

belong to a regulus. By the matrix cross-ratio theorem this reduces to the condition that the matrix

$$[(\alpha_1+\alpha_2)^2 I - (T_1 - T_2)^2]/(4\alpha_1\alpha_2)$$

should be a scalar multiple of I; whence, since $T_1^2 = t_1 I$ and $T_2^2 = t_2 I$, it follows that $T_1 T_2 + T_2 T_1 = 2t_{12} I$, where t_{12} is a scalar.

LEMMA D. *If $y = A_1 x$ and $y = A_2 x$ are two independent generic spaces of a linear system $\mathscr{F}$ which contains R_0, then $\mathscr{F}$ contains also the space $y = (\mu_1 A_1 + \mu_2 A_2)x$ for arbitrary μ_1, μ_2.*

Proof. Since $\mathscr{F}$ is irreducible, it is enough to prove that the space in question belongs to $\mathscr{F}$ when μ_1 and μ_2 are taken to be independent indeterminates over the field obtained by adjoining the elements of A_1 and A_2 to the ground field. By Lemmas A and C we know that

$$A_1 = \alpha_1 I + T_1 \quad \text{and} \quad A_2 = \alpha_2 I + T_2,$$

where $\alpha_1 \alpha_2 \neq 0$ and the skew-symmetric matrices T_1, T_2 satisfy

$$T_1^2 = t_1 I, \quad T_2^2 = t_2 I, \quad T_1 T_2 + T_2 T_1 = 2t_{12} I$$

for suitable t_1, t_2, t_{12}. If we write

$$\mu_1' = \frac{\mu_1}{\mu_1 + \mu_2}, \quad \mu_2' = \frac{\mu_2}{\mu_1 + \mu_2}$$

and take ϕ to be either root of the equation

$$\phi^2 = \mu_1' \alpha_1^2 + \mu_2' \alpha_2^2 - \mu_1' \mu_2' (t_1 + t_2 - 2t_{12}),$$

then $\phi \neq 0$ and we find, with little difficulty, that the space

$$y = (\phi I + \mu_1' T_1 + \mu_2' T_2)\, x$$

is a member of the Clifford regulus determined by the spaces

$$y = (\alpha_1 I + T_1)\, x \quad \text{and} \quad y = (\alpha_2 I + T_2)\, x$$

and so belongs to $\mathscr{F}$. (This amounts to a verification, by the matrix cross-ratio theorem, that the three spaces in question, together with the polar space $y = (-\alpha_1 I + T_1)\, x$ of one of the last two, belong to a regulus.) Hence, by Lemma B, the space

$$y = [\delta I + \epsilon(\mu_1' T_1 + \mu_2' T_2)]\, x$$

belongs to $\mathscr{F}$ for all δ, ϵ. In particular, taking $\delta = \alpha_1 \mu_1 + \alpha_2 \mu_2$ and $\epsilon = \mu_1 + \mu_2$, it follows that $\mathscr{F}$ contains the space

$$y = [(\alpha_1 \mu_1 + \alpha_2 \mu_2)\, I + \mu_1 T_1 + \mu_2 T_2]\, x = (\mu_1 A_1 + \mu_2 A_2)\, x$$

as required.

Proof of Theorem 3.1. We are now in a position to complete the proof of the first part of the main theorem. We may suppose, as already noted, that the linear system $\mathscr{F}$ of Clifford parallels contains the regulus R_0 with equation $y = \lambda x$. Then almost all members of $\mathscr{F}$ have equations of the form $y = Ax$; and it follows now from Lemma D that the matrices A for which $y = Ax$ belongs to $\mathscr{F}$ form a vector subspace $\mathscr{V}$ of the vector space of all matrices of order n, and the dimension of $\mathscr{V}$ is equal to the dimension r of $\mathscr{F}$. Further, with our choice of coordinate system, it follows that $\mathscr{V}$ contains all scalar multiples of I; and, by Lemma A, every element of $\mathscr{V}$ is of the form $\alpha I + T$, where T is skew-symmetric and such that T^2 is a scalar multiple of I. By Lemmas A and C, therefore, we may choose a basis of linearly independent matrices

$$I, T_1, \ldots, T_{r-1}$$

for $\mathscr{V}$ such that $T_1, \ldots, T_{r-1}$ are skew-symmetric and

$$T_i T_j + T_j T_i = 2t_{ij} I$$

for all i, j (distinct or not), where the t_{ij} are scalars.

To simplify this further, we modify the basis as follows. Let $\mathscr{T}$ denote the matrix (t_{ij}) which is symmetric of order $r-1$ and let the rank of $\mathscr{T}$ be $k-1$. There exists then a non-singular matrix $M = (m_{ij})$, of order $r-1$, such that

$$M \mathscr{T} M^T = -D,$$

where $D = (d_{ij})$ is the diagonal matrix $[1^{k-1}, 0^{r-k}]$. If we now write

$$S_i = \sum_{j=1}^{r-1} m_{ij} T_j \quad (i = 1, \ldots, r-1),$$

then I, $S_1, \ldots, S_{r-1}$ is another basis for $\mathscr{V}$ (with the S_i skew-symmetric); and an easy calculation shows that

$$S_i S_j + S_j S_i = -2d_{ij} I$$

for all i, j. This completes the proof of the first part of Theorem 3.1.

As regards the basis I, $S_1, \ldots, S_{r-1}$ just chosen for $\mathscr{V}$, we note that this is by no means uniquely determined. Thus, if L is any orthogonal matrix of order $k-1$, if M is any non-singular matrix of order $r-k$, and if N is an arbitrary matrix of $k-1$ rows and $r-k$ columns, then the matrix

$$C = (c_{ij}) = \begin{pmatrix} L & N \\ 0 & M \end{pmatrix}$$

has the property $CDC^T = D$, and every C satisfying this equation is of the form indicated. Hence the matrices $S'_1, \ldots, S'_{r-1}$ given by

$$S'_i = \sum_{j=1}^{r-1} c_{ij} S_j \quad (i = 1, \ldots, r-1)$$

satisfy effectively the same equations as $S_1, \ldots, S_{r-1}$.

Turning now to the second (sufficiency) part of Theorem 3.1, we have to prove that, if $S_1, \ldots, S_{r-1}$ satisfy the equations (3.1) for some k $(1 \leqslant k \leqslant r)$, then a generic pair of spaces of the system given by (3.2) are Clifford parallel, and that the Clifford regulus joining them is contained in the system. We prove this directly as follows.

Let Π_1 and Π_2 be the generic pair in question, with respective equations

$$y = \left(\alpha I + \sum_{i=1}^{r-1} \lambda_i S_i\right) x,$$

and

$$y = \left(\beta I + \sum_{i=1}^{r-1} \mu_i S_i\right) x,$$

where the $2r$ symbols $\alpha, \beta, \lambda_1, \ldots, \lambda_{r-1}, \mu_1, \ldots, \mu_{r-1}$ denote independent indeterminates. The polar space Π_1' of Π_1 then has equation

$$y = \left(-\alpha I + \sum_{i=1}^{r-1} \lambda_i S_i\right) x.$$

If we make the allowable coordinate transformation

$$\xi = y - \left(-\alpha I + \sum_{i=1}^{r-1} \lambda_i S_i\right) x, \quad \eta = iy - i\left(\alpha I + \sum_{i=1}^{r-1} \lambda_i S_i\right) x,$$

so that Π_1 and Π_1' are now given by $\eta = 0$ and $\xi = 0$ respectively, then the new equation of Ω reduces, by use of equations (3.1), to

$$\xi^T \xi + \eta^T \eta = 0$$

and, if we write

$$\psi = \frac{1}{2\alpha}\left\{(\alpha+\beta)^2 + \sum_{i=1}^{k-1} (\mu_i - \lambda_i)^2\right\},$$

we find with little difficulty that the new equation of Π_2 can be written in the form $\eta = C\xi$, where

$$C = i\psi^{-1}\left\{(\psi - \alpha - \beta) I + \sum_{i=1}^{r-1} (\mu_i - \lambda_i) S_i\right\}.$$

A simple calculation reveals moreover that $C^T C = (2\beta\psi^{-1} - 1) I$. Hence, since $2\beta\psi^{-1} - 1 \neq 0$ or -1, it follows from Prop. 2.1 of Ch. 3 that $\eta = C\xi$ is Clifford parallel to $\eta = 0$ with respect to $\xi^T\xi + \eta^T\eta = 0$, i.e. that Π_2 is Clifford parallel to Π_1 with respect to Ω. Further (cf. Ch. 3, § 2), the equation of the Clifford regulus joining Π_2 to Π_1 is $\eta = \theta C\xi$, where θ is a variable parameter; and, converting back to coordinates (x, y), it is easily verified that the spaces of this regulus all belong to the system defined by equation (3.2). The proof of Theorem 3.1 is thereby complete.†

If we premultiply equation (3.2) by x^T, we obtain

$$x^T y = \alpha x^T x,$$

which shows that, if $\alpha \neq 0$, the space Π given by (3.2) is met in the same quadric by Ω and by the cone $x^T x = 0$. From this, and by using the results in Ch. 3, § 5, we deduce

COROLLARY 3.1.1. *If equation* (3.2) *represents a generator* Π *of* Ω (*i.e. if* $\alpha = 0$), *then* Π *has the same terminal quadric* q (*given by* $x^T x = 0$) *in every Clifford regulus of* $\mathscr{F}$ *of which it is a terminal generator.*

The significance of this result is that Π, by virtue solely of its membership of $\mathscr{F}$, acquires the structure of a unique augmentation $\Pi(q)$ of Π; and we shall express this later (Ch. 6, § 5) by saying that this $\Pi(q)$ *properly belongs* to $\mathscr{F}$.

Finally, in this section, we note that if (for given values of n, r and k) $S_1, \ldots, S_{r-1}$ is a set of matrices satisfying (3.1), and if Q is any (complex) orthogonal matrix of order n, then

$$Q^T S_1 Q, \ldots, Q^T S_{r-1} Q$$

is likewise a set of matrices satisfying (3.1) (for the same values of n, r and k) and gives rise, therefore, to a linear system $\mathscr{F}_Q$ of Clifford parallels with equation

$$y = (\alpha I + \lambda_1 Q^T S_1 Q + \ldots + \lambda_{r-1} Q^T S_{r-1} Q)\, x. \tag{3.4}$$

If we then make the allowable coordinate transformation

$$\xi = Qx, \quad \eta = Qy,$$

we see that (3.4) reduces to

$$\eta = (\alpha I + \lambda_1 S_1 + \ldots + \lambda_{r-1} S_{r-1})\, \xi,$$

† The algebra of the above proof can be used to show that the Clifford reguli in $\mathscr{F}$ are represented on the affine space of $\alpha, \lambda_1, \ldots, \lambda_{r-1}$ by conics (of a certain family) having $\alpha = 0$ as a prime of symmetry; but a simpler and more informative derivation of this result comes from the representational theory of Ch. 6 (cf. Ch. 6, § 2, Ex. 3).

while the equation of Ω retains the canonical form $\xi^T\eta = 0$. It follows, therefore, that the system $\mathscr{F}_Q$ is projectively equivalent (by an Ω-preserving collineation) to the system $\mathscr{F}$ of Theorem 3.1. Hence we may record

COROLLARY 3.1.2. *In order to find all projectively distinct linear systems of Clifford parallels (for given n, r and k), it is sufficient to find all the solution-sets of equations* (3.1) *up to orthogonal equivalence.*

It is possible, however, for two solution-sets to give rise to projectively equivalent linear systems of Clifford parallels, even when they are not orthogonally equivalent.† But, in the interesting cases at least, the simplification afforded by Cor. 3.1.2 is sufficient to reduce the problem of classifying linear systems of Clifford parallels to manageable proportions (cf. § 5 below).

Ex. 1. For $n = 2$, every skew-symmetric matrix is a scalar multiple of

$$S_1 = \begin{pmatrix} 0 & -1 \\ 1 & 0 \end{pmatrix},$$

for which $S_1^2 = -I$. By Th. 3.1, then, we obtain a 2-dimensional linear system of Clifford parallels in S_3 with equation

$$\begin{pmatrix} y_0 \\ y_1 \end{pmatrix} = \begin{pmatrix} \alpha & \lambda_1 \\ -\lambda_1 & \alpha \end{pmatrix} \begin{pmatrix} x_0 \\ x_1 \end{pmatrix}$$

and this is the only 2-dimensional system that can exist (up to projective equivalence). The lines of the system form a linear congruence, being all the transversals of the two skew generators of Ω given by $x_1 = \pm ix_0$, $y_1 = \pm iy_0$ (cf. Ch. 3, § 2, Ex. 2).

For $n = 3$, there exists no non-zero skew-symmetric matrix S such that S^2 is either $-I$ or 0; and there are therefore no linear systems of Clifford parallel planes of S_5 that are more ample than the Clifford regulus.

Ex. 2. If $S_1, \ldots, S_{r-1}$ are skew-symmetric matrices, anti-commuting by pairs, such that $S_1^2 = \ldots = S_{r-1}^2 = -I$, show that

† If $(S_1, \ldots, S_{r-1})$ is a solution-set of (3.1), then so also is $(\pm S_1, \ldots, \pm S_{r-1})$, with arbitrary sign combinations, and the latter is not always orthogonally equivalent to the former. But the two solutions plainly give rise to one and the same linear system of Clifford parallels.

$S_1, \ldots, S_{r-1}$ are necessarily linearly independent. (The same conclusion does not follow from equations (3.1) if $k < r$.)

Ex. 3. If $S_1, \ldots, S_{r-1}$ satisfy equations (3.1), show that each product $S_i S_j$ $(i \neq j)$ is a skew-symmetric matrix.

Ex. 4. If $k > 1$ in the equations (3.1), show that n is necessarily even. (N.B. A skew-symmetric matrix of odd order is singular.)

Ex. 5. Further to the proof of the second part of Th. 3.1, establish the following slightly stronger result: *every* space of $\mathscr{F}$ given by the equation (3.2) belongs to some Clifford regulus within $\mathscr{F}$.

4. Non-singular and singular linear systems

If Ω has equation $x^T y = 0$ and if R_0 is the Clifford regulus $y = \theta x$, then Th. 3.1 tells us that any linear ∞^r-system $\mathscr{F}$ of Clifford parallels which contains R_0 has an equation of the form

$$y = (\alpha I + \lambda_1 S_1 + \ldots + \lambda_{r-1} S_{r-1})\, x,$$

where $S_1, \ldots, S_{r-1}$ are $r-1$ linearly independent matrices of order n satisfying the extended Hurwitz–Radon equations (3.1) for some value of k in the range $1 \leqslant k \leqslant r$, and this equation covers all the spaces of $\mathscr{F}$ that do not meet the reference space Y. For brevity we shall write

$$S(\lambda) = \lambda_1 S_1 + \ldots + \lambda_{r-1} S_{r-1},$$

so that the equation of $\mathscr{F}$ takes the form

$$y = (\alpha I + S(\lambda))\, x. \tag{4.1}$$

We note, by virtue of equations (3.1), that

$$(S(\lambda))^2 = \left(\sum_1^{r-1} \lambda_i S_i\right)^2 = \Lambda^2 I,$$

where $\Lambda^2 = -(\lambda_1^2 + \ldots + \lambda_{k-1}^2)$ if $k > 1$, and $\Lambda = 0$ if $k = 1$. We now make the following observations.

(i) If Π is the space (4.1), then its polar space Π' is given by

$$y = (-\alpha I + S(\lambda))\, x.$$

(ii) If $\alpha = 0$, then Π is a generator of Ω.

(iii) If $\alpha \neq 0$, then Π is skew to Π', i.e. Π is chordal.

At this point we must avoid the error of assuming, by virtue of (ii) and (iii) above, that every space of $\mathscr{F}$ must be either chordal or a generator of Ω; for (ii) and (iii) apply only to spaces of $\mathscr{F}$ with equations of the form (4.1), and this excludes spaces of $\mathscr{F}$ which meet the reference space Y. To simplify further discussion of this point we introduce the

DEFINITION. If $\mathscr{F}$ is a linear system of Clifford parallels and if Σ is a terminal generator of a Clifford regulus contained in $\mathscr{F}$, then the spaces of $\mathscr{F}$ which do not meet Σ will be said to form the Σ-*affine part* of $\mathscr{F}$.

With this terminology we may say that equation (4.1) represents explicitly all the spaces of the Y-affine part of $\mathscr{F}$, and that every space of this Y-affine part is either chordal or a generator of Ω. More generally, allowing for the arbitrary choice of Y in the formulation of Th. 3.1, we have

PROPOSITION 4.1. *If* Σ *is any terminal generator of a Clifford regulus in the linear system* $\mathscr{F}$, *then every space in the* Σ-*affine part of* $\mathscr{F}$ *is either chordal or a generator of* Ω.

We now introduce the further

DEFINITION. If a space of $\mathscr{F}$ is such that it does not belong to the Σ-affine part of $\mathscr{F}$ for any admissible Σ, then it will be called a *singular space* of $\mathscr{F}$.

It follows then, by Prop. 4.1, that a space of $\mathscr{F}$ that is neither chordal nor a generator of Ω is necessarily a singular space of $\mathscr{F}$. Further, a singular space of $\mathscr{F}$ must meet both terminal generators of every Clifford regulus in $\mathscr{F}$; and, in particular, it cannot belong to any Clifford regulus in $\mathscr{F}$. Conversely, since every space of the Σ-affine part of $\mathscr{F}$ belongs to Clifford reguli contained in $\mathscr{F}$ (cf. § 3, Ex. 5) we have

COROLLARY 4.1.1. *A space of* $\mathscr{F}$ *is singular if and only if it does not belong to any Clifford regulus in* $\mathscr{F}$.

Since singular spaces, by definition, can never be represented explicitly in any canonical equation of $\mathscr{F}$, they are naturally difficult to locate; but examples can be given to show that they do sometimes exist (cf. Ex. 4 below).

We now remark, however, that there are good reasons for distinguishing generally between those linear systems which do not contain singular spaces and those which do. While we are not yet in a position to explain these reasons fully, it may assist the reader at this stage to be given some insight into the general situation. In brief, the key to this is the value of the integer k that occurs in the statement of Th. 3.1. So far we have not proved that k is an invariant† of the linear system $\mathscr{F}$, although this is indeed the case. A direct proof of this, at the present stage of development is straightforward but tedious (involving an extension of the algebra developed in proving the second half of Th. 3.1), and the reader may care to supply his own proof. We prefer, however, to defer this point until Ch. 6, Cor. 2.2.1, where it emerges that k has a natural geometrical meaning relative to $\mathscr{F}$ and so does not depend on any algebraic presentation of the equations of $\mathscr{F}$. Granting then the invariance of k, we introduce the

DEFINITION. If the equation of a linear ∞^r-system $\mathscr{F}$ is given in the canonical form of Th. 3.1, then the number k so arising will be called the *weight* of the system $\mathscr{F}$. If $k = r$, we shall say that $\mathscr{F}$ is *non-singular*; if $k < r$, then $\mathscr{F}$ is *singular*; and in the extreme case (in some ways abnormal) in which $k = 1$, we shall say that $\mathscr{F}$ is *totally singular*.

In terms of this definition, the situation as regards the possible existence of singular spaces of $\mathscr{F}$ may be stated as follows: *a linear system $\mathscr{F}$ possesses singular spaces if and only if it is singular*. The justification of this statement will appear in Ch. 6, §5.

If we take the view, as in general we do, that the most interesting systems of mutually Clifford parallel spaces are those

† To show directly that k is an invariant of $\mathscr{F}$, it would be necessary to prove, in connection with the proof of Th. 3.1, that k is independent of the choice of R_0 from among the Clifford reguli contained in $\mathscr{F}$, as also of the choice of the basis $I, S_1, \ldots, S_{r-1}$ for the vector space $\mathscr{V}$ of matrices associated with $\mathscr{F}$.

which conform reasonably to our intuitive ideas of parallelism, it is found that singular systems can present anomalies. Thus, for example, a singular system may contain pairs of spaces which have common points not 'at infinity', i.e. not on Ω; and we give an example of this below (Ex. 4). Or again, a singular system of Clifford parallel $[n-1]$'s of S_{2n-1} can be such that it is of dimension $r > n$, in which case it must certainly contain pairs of spaces with common points not at infinity; and we give an example of this also (Ex. 5 below). For non-singular systems, however, these anomalies do not arise; and in fact *if two distinct spaces of a non-singular linear system of Clifford parallels intersect, then their intersection lies on* Ω. Here again the proof of this statement must be deferred till we have developed the representational method of Ch. 6; but at this stage we can prove the following weaker result:

PROPOSITION 4.2. *If* Π_1 *and* Π_2 *are distinct spaces belonging to the* Σ*-affine part of a non-singular linear system* $\mathscr{F}$ *for some admissible* Σ, *and if* Π_1 *and* Π_2 *intersect, then their intersection lies on* Ω.

Proof. Without loss of generality we may take Σ to be the reference space Y, so that the spaces of the Σ-affine part of $\mathscr{F}$ may be taken to be those of the system given by the canonical equation (3.2), in which $S_1, \ldots, S_{r-1}$ satisfy equations (3.1) with $k = r$. We may assume, of course, that neither Π_1 nor Π_2 is a generator of Ω, and hence that Π_1 and Π_2 meet Ω in the quadrics q_1 and q_2 in which they are met by the cone $x^T x = 0$ (cf. the remarks preceding Cor. 3.1.1).

Consider then a point (ξ, η) which is such that $\xi^T \xi \neq 0$ and which lies therefore in neither of q_1, q_2. Any space of the Σ-affine part of $\mathscr{F}$ that passes through (ξ, η) – if any such space exists – will be given by values of $\alpha, \lambda_1, \ldots, \lambda_{r-1}$ that satisfy

$$\eta = (\alpha I + \lambda_1 S_1 + \ldots + \lambda_{r-1} S_{r-1})\,\xi. \tag{4.2}$$

If we premultiply (4.2) by $\xi^T, \xi^T S_1, \ldots, \xi^T S_{r-1}$ in turn, we find (using Ex. 3 of § 3) that

$$\xi^T \eta = \alpha \xi^T \xi \quad \text{and} \quad \xi^T S_i \eta = -\lambda_i \xi^T \xi; \tag{4.3}$$

and this means, since $\xi^T\xi \neq 0$, that if there exist values of $\alpha, \lambda_1, \ldots, \lambda_{r-1}$ satisfying (4.2), then these values are uniquely determined by (4.3). Hence any point common to the two (distinct) spaces Π_1, Π_2 must satisfy the equation $x^Tx = 0$, i.e. it must be a point of Ω (common to q_1, q_2). This proves the Proposition.

A further remark arises from the above analysis. If we substitute the values of α, $\lambda_1, \ldots, \lambda_{r-1}$ given by (4.3) back into (4.2) and replace (ξ, η) by the current coordinates (x, y), we obtain the equation

$$yx^Tx = xx^Ty - S_1xx^TS_1y - \ldots - S_{r-1}xx^TS_{r-1}y \qquad (4.4)$$

which is satisfied by every point (ξ, η) that lies in a space of the Σ-affine part of $\mathscr{F}$; also, conversely, any point (ξ, η) that satisfies (4.4) and is such that $\xi^T\xi \neq 0$ lies in some space of the Σ-affine part of $\mathscr{F}$. Equation (4.4), equivalent to n ordinary equations, represents a variety in S_{2n-1} which is in general reducible, consisting in part of the variety V generated by all the spaces of $\mathscr{F}$, and residually of one or more components contained in the cone $x^Tx = 0$. The dimension of V is $n-1+r$; and, in particular, if $r = n$, then V must be the whole space S_{2n-1} (equation (4.4) then reducing to an identity). This gives

COROLLARY 4.2.1. *If $n = r = k$ in the canonical equation* (3.2) *of a linear system $\mathscr{F}$, then the spaces of $\mathscr{F}$ form a space-filling system of index* 1.

EX. 1. Using the identity

$$(\alpha I + S(\lambda))\,(\alpha I - S(\lambda)) = (\alpha^2 - \Lambda^2)\,I,$$

where $\Lambda^2 = -(\lambda_1^2 + \ldots + \lambda_{k-1}^2)$, show that the system $\mathscr{F}$ given by equation (4.1) admits the alternative representation

$$x = (\beta I + \mu_1 S_1 + \ldots + \mu_{r-1}S_{r-1})\,y,$$

where β, $\mu_1, \ldots, \mu_{r-1}$ are arbitrary parameters. This equation represents explicitly all the spaces of $\mathscr{F}$ that do not meet the reference space X, i.e. the X-affine part of $\mathscr{F}$. (N.B. The condition for (4.1) to meet X is $\alpha^2 - \Lambda^2 = 0$.)

EX. 2. If $\mathscr{F}$ is a linear system, with generic member Π given by (3.2), and of weight $k > 1$ (so that $S_1^2 = -I$ in (3.1) and n is even), show that Π meets two generators of the Clifford regulus R_0 given by $y = \theta x$ and that it meets each of them in a space $[\frac{1}{2}n-1]$ which is contained in the kernel of R_0.

[Hint: If $A = (\alpha-\theta)I+S(\lambda)$ then $AA^T = \{(\alpha-\theta)^2-\Lambda^2\}I$, where $\Lambda^2 \neq 0$. Hence, by well-known rank inequalities, A is of rank $\frac{1}{2}n$ when $\theta = \alpha \pm \Lambda$, and the equations of the two $[\frac{1}{2}n-1]$'s in question are

$$y = (\alpha \pm \Lambda)x, \quad \{\Lambda I \mp S(\lambda)\}x = 0.]$$

Important deductions from this result will be made in Ch. 5, § 3.

Ex. 3. If $\mathscr{F}_1$ and $\mathscr{F}_2$ are two linear systems, of weights k_1 and k_2, such that $\mathscr{F}_1 \subset \mathscr{F}_2$, show that $k_1 \leqslant k_2$.

Ex. 4. For $n = 5$, let S_1 denote the skew-symmetric matrix

$$\begin{pmatrix} 0 & 1 & 0 & i & 0 \\ -1 & 0 & i & 0 & 0 \\ 0 & -i & 0 & 1 & 0 \\ -i & 0 & -1 & 0 & 0 \\ 0 & 0 & 0 & 0 & 0 \end{pmatrix}.$$

Verify that $S_1^2 = 0$ and deduce that the ∞^2-system of [4]'s of S_9 given by $y = (\alpha I+\lambda_1 S_1)x$ is a (totally singular) linear system $\mathscr{F}$ of Clifford parallels relative to the quadric $x^Ty = 0$.

By letting λ_1 tend to ∞ (keeping α finite and non-zero), show that this system $\mathscr{F}$ contains singular spaces, i.e. [4]'s which are neither chordal nor generators of Ω.

Show further that those [4]'s of $\mathscr{F}$ for which α has a given value all pass through a fixed plane which does not lie on Ω.

Ex. 5. Let $n = 2m$, let T be a skew-symmetric matrix of order m whose elements t_{ij} above the diagonal are independent indeterminates, let β be a further indeterminate, and let S be the matrix given by

$$S = \begin{pmatrix} T & -\beta I+iT \\ \beta I+iT & -T \end{pmatrix}.$$

If S' denotes the matrix obtained from S by replacing the t_{ij} and β by t'_{ij} and β', verify that

$$SS'+S'S = -2\beta\beta' I$$

and deduce that $y = (\alpha I + S)x$ is the generic space of a linear system of Clifford parallels (relative to $x^T y = 0$) of freedom $\frac{1}{8}n(n-2)+2$. [If n is even and $n \geqslant 10$, this implies that there exist (singular) linear systems of Clifford parallel $[n-1]$'s of S_{2n-1} of freedom $r > n$.] The system is of weight $k = 2$ and, for reasons that appear later (Ch. 5, § 3, Ex. 7) we call it a *half-fixed system.*

If we split the column vectors x and y into two halves, writing the equation of the system in the form

$$\begin{pmatrix} y_1 \\ y_2 \end{pmatrix} = \begin{pmatrix} \alpha I + T & -\beta I + iT \\ \beta I + iT & \alpha I - T \end{pmatrix} \begin{pmatrix} x_1 \\ x_2 \end{pmatrix},$$

show that this equation implies the relations:

$$y_1 + iy_2 = (\alpha + i\beta)(x_1 + ix_2),$$

$$x_1^T y_1 + x_2^T y_2 = \alpha(x_1^T x_1 + x_2^T x_2),$$

$$x_1^T y_2 - x_2^T y_1 = \beta(x_1^T x_1 + x_2^T x_2).$$

Deduce that the spaces of the half-fixed system do not fill up S_{2n-1}, and find equations for the locus which they generate.

5. The classification problem

The problem of classifying linear systems of Clifford parallels is equivalent, as we have seen (Cor. 3.1.2) to finding all solution-sets of the extended Hurwitz–Radon matrix equations (3.1) (for given n, r and k) up to orthogonal equivalence; and it would be interesting to know the possible values of n, r and k for which *maximal linear systems* exist (i.e. systems which are not contained in any more ample linear systems). This would appear, however, to be a problem of considerable difficulty and we shall not attempt to solve it in this book.

As we have already indicated, however, there are good reasons why we should be specially interested in *non-singular* linear systems and, for this case at least, the solution of the classification problem can be given. In fact, when $k = r$, equations (3.1) reduce to the well-known and extensively documented Hurwitz–Radon matrix equations and we refer the reader to the Appendix for details. The main result (Appendix, Th. A1) is as follows:

If n is written in the form $n = u.\,2^{4\alpha+\beta}$, *with u odd and* $0 \leqslant \beta \leqslant 3$, *then equations* (3.1) – *with* $k = r$ – *admit solutions in matrices of order n if and only if* $r \leqslant H(n)$, *where* $H(n) = 8\alpha + 2^{\beta}$. *Moreover there are, up to orthogonal equivalence, only a finite number of solution-sets for given n and r so that, by Cor.* 3.1.2, *there are only a finite number of projectively distinct non-singular linear systems of Clifford parallels in* S_{2n-1}.

The values of $H(n)$ corresponding to low values of n are those given in the table

n	2	3	4	5	6	7	8	9	10	11	12
$H(n)$	2	1	4	1	2	1	8	1	2	1	4

We note also that $H(n) \leqslant n$ for all n. Furthermore, the equality $H(n) = n$ occurs only for $n = 2$, 4 or 8; so that there can (and in fact do) exist *space-filling* non-singular linear systems of Clifford parallels for only these three values of n, i.e. in S_3, S_7 and S_{15}. We shall give much attention in subsequent chapters to explicit geometrical constructions for these space-filling systems (Ch. 5) and to methods of representing them (Ch. 6 and Ch. 7).

Ex. 1. The formula quoted above always gives $H(n) = 1$ when n is odd. This means that, for n odd, there are no non-singular linear systems more ample than the Clifford regulus. Confirm this result by using Ex. 4 of § 3.

Ex. 2. If $n = 8$, $H(n) = 8$ and there accordingly exists a non-singular linear ∞^8-system of Clifford parallel [7]'s in S_{15}. There also exists, however, a non-singular linear ∞^4-system which is *maximal* and hence not embeddable in any ∞^8-system (cf. Ch. 5, § 4).

This illustrates the general point that, if a solution-set of the Hurwitz–Radon equations contains fewer than $H(n)$ matrices, then it is not necessarily possible to augment the solution-set to one containing $H(n)$ matrices.

6. Linear systems $\mathscr{F}$ related to a Clifford kernel

If we are given in S_{2n-1} the absolute quadric Ω and any one Clifford regulus R_0 with respect to Ω, then we know by Th. 3.1 that the set $(\mathscr{F})$ of all linear systems of Clifford parallels with respect to Ω that contain R_0 is determined by the set of all solutions, for all possible values of r and k, of the extended Hurwitz–Radon matrix equations (3.1). We now propose to show that the set $(\mathscr{F})$ depends only on the kernel K_0 of R_0 with respect to Ω; more precisely, that $(\mathscr{F})$ is equally the set of linear systems of Clifford parallels containing R_0 with respect to every non-singular quadric Ω' of a certain net uniquely defined by K_0.

PROPOSITION 6.1. *If Ω has equation $x^T y = 0$, R_0 is the regulus $y = \theta x$, and $\mathscr{F}$ (containing R_0) is the linear system of Clifford parallels with respect to Ω given by equation* (3.2), *then $\mathscr{F}$ is also a linear system of Clifford parallels with respect to every non-singular quadric of the net*

$$ax^T x + 2bx^T y + cy^T y = 0. \tag{6.1}$$

Proof. For the purposes of the proof we regard $\alpha, \lambda_1, \ldots, \lambda_{r-1}$ in (3.2) as independent indeterminates and we write

$$S = \sum_{i=1}^{r-1} \lambda_i S_i \quad \text{and} \quad \Lambda^2 = -(\lambda_1^2 + \ldots + \lambda_{k-1}^2),$$

so that $S^2 = \Lambda^2 I$.

Suppose then that θ_1, θ_2 are the roots of the equation

$$a + 2b\theta + c\theta^2 = 0$$

(which are distinct if and only if $b^2 \neq ac$, i.e. if and only if (6.1) represents a non-singular quadric) and let (ξ, η) be new coordinates in S_{2n-1} related to the coordinates (x, y) by the allowable transformation

$$x = \theta_1 \xi - \eta, \quad y = \theta_2 \xi - \eta.$$

The new equation of the general space $y = \theta x$ of R_0 is then

$$\eta = \phi \xi, \quad \text{where} \quad \phi = \frac{\theta\theta_1 - \theta_2}{\theta - 1},$$

the new equation of $x^T y = 0$ is

$$a\xi^T \xi + 2b\xi^T \eta + c\eta^T \eta = 0, \tag{6.2}$$

and the new equation of $\mathscr{F}$ is

$$\theta_2 \xi - \eta = (\alpha I + S)(\theta_1 \xi - \eta),$$

which can be rearranged in the form

$$\begin{aligned} \eta &= \{(\alpha-1)I + S\}^{-1}\{(\alpha\theta_1 - \theta_2)I + \theta_1 S\}\xi \\ &= \{(\alpha-1)^2 - \Lambda^2)\}^{-1}\{(\alpha-1)I - S\}\{(\alpha\theta_1 - \theta_2)I + \theta_1 S\}\xi \\ &= \{(\alpha-1)^2 - \Lambda^2\}^{-1}\{[(\alpha-1)(\alpha\theta_1 - \theta_2) - \Lambda^2\theta_1]I + (\theta_2 - \theta_1)S\}\xi; \end{aligned}$$

and, since $\alpha, \lambda_1, \ldots, \lambda_{r-1}$ are independent indeterminates, this is equivalent to

$$\eta = (\beta I + \mu_1 S_1 + \ldots + \mu_{r-1} S_{r-1})\xi \tag{6.3}$$

where $\beta, \mu_1, \ldots, \mu_{r-1}$ are new independent indeterminates.

It follows now that $\mathscr{F}$, in the form (6.3), is a linear system of Clifford parallels with respect to the quadric (6.2) for arbitrary values of a, b, c such that $b^2 \neq ac$, and this proves Prop. 6.1.

In amplification of the preceding Proposition, we add the following Corollary, whose proof may be left to the reader.

COROLLARY 6.1.1. *If K_0 is the kernel of the Clifford regulus R_0 (with equation $y = \theta x$) relative to $x^T y = 0$, then every quadric containing K_0 is a member of the system given by*

$$ax^T x + 2bx^T y + cy^T y + x^T S y = 0$$

where S is skew-symmetric; but R_0 is a Clifford regulus only for those non-singular quadrics of this system for which $S = 0$. In other words, when a Clifford regulus and its Clifford kernel are given, the absolute quadric is determined to the extent that it must belong to a certain net.

Finally we prove

PROPOSITION 6.2. *If (Ω) is the net of quadrics uniquely associated, as in Cor. 6.1.1, with a kernel K_0 on R_0, then for $n \geqslant 4$ the base variety of (Ω) is the V^8_{2n-4} generated by the ∞^{2n-7} solids (not lying in generators of R_0) which meet K_0 in quadric surfaces.*

Proof. We note first that the quadric given by (6.1) is a cone if and only if $b^2 = ac$, and that the equation of any such cone is of the form $(y-\psi x)^T(y-\psi x) = 0$ for some value of ψ. This equation represents the cone projecting K_0 from the generator $y = \psi x$ of R_0. Thus (Ω) is the net of quadrics which contains all the $[n-1]$-cones projecting K_0 from generators of R_0.

Consider then a point P, not on R_0, which lies on all the cones in question. Let Π_1, Π_2 be any pair of generators of R_0 meeting K_0 in quadrics q_1, q_2; and let t be the unique transversal line from P to Π_1, Π_2. Since P lies on the cones $\Pi_1(q_2)$ and $\Pi_2(q_1)$, it follows that t meets Π_1 in a point of q_1 and Π_2 in a point of q_2. Thus P is such that the chords of R_0 which pass through it are all chords of K_0. This means again that the unique quadric solid of R_0 through P is in fact a quadric solid of K_0; and the base locus of (Ω) is therefore generated by quadric solids of K_0.

Now K_0, being the Segre product of an $(n-2)$-dimensional quadric q by a line l, possesses ∞^{2n-7} quadric surfaces (not lying in generators of R_0), each the product of l by a line on q; and the required octavic base locus of the net (Ω) is therefore the V^8_{2n-4} generated by the solids which meet K_0 in those quadric surfaces. This completes the proof.

Note. The cases $n = 2$ and $n = 3$ are exceptional because K_0 does not possess quadric solids for these values of n. For $n = 2$, the base of the net (Ω) is K_0 itself – a pair of skew lines; and for $n = 3$ the base is K_0 counted twice, K_0 being a rational ruled surface ${}^0R^4_2$.

The case $n = 4$ is special in another way. For in this case K_0, being the Segre product of a quadric surface q by a line l, is also a *Segre product of three lines* (since q itself is a product of two lines); and it therefore possesses two ∞^1-systems of quadric-solids in addition to the systems of generating solids of R_0, i.e. three such systems in all, generating three reguli of which R_0 is one. (We shall return to this later in Ch. 5, § 2, Ex. 3.) Thus for $n = 4$ the base variety V^8_4 of (Ω) breaks up into two V^4_4's – the two reguli other than R_0 associated with K_0 in this case.

For $n > 4$, the lines on a quadric q^2_{n-2} form a single irreducible system, so that the base V^8_{2n-4} of (Ω) is irreducible.

Ex. 1. From Prop. 6.1 it appears in particular that, if Ω has equation $x^T x + y^T y = 0$, then any linear system $\mathscr{F}$ of Clifford parallels which contains the Clifford regulus $y = \theta x$ has a canonical equation of the form given in Th. 3.1. This canonical form (for non-singular systems in elliptic space EL_{2n-1}) was originally established by Wong [**27**] with the equation of Ω in the above form.

Ex. 2. Discuss the character of the net (Ω) which is given for $n = 3$ by the equation

$$a(x_0^2 + x_1^2 + x_2^2) + 2b(x_0 y_0 + x_1 y_1 + x_2 y_2) + c(y_0^2 + y_1^2 + y_2^2) = 0.$$

CHAPTER 5

GEOMETRICAL CONSTRUCTIONS

In the previous Chapter, our main concern was with the formulation and discussion of an existence problem, namely that of the existence of linear systems – non-singular or singular – of Clifford parallel $[n-1]$'s of S_{2n-1}; and in § 3 of the same Chapter we reduced this problem to its equivalent in terms of matrix equations of which the solutions for the non-singular case were already known. In the present Chapter we turn to more geometrical matters, in particular to the problem of finding explicit geometrical constructions for the linear systems in question. The general construction problem that arises in this connection is one of considerable complexity to which we do not at present know all the answers. It would seem useful, however, to give the details of its solution at least in those cases ($n = 2, 4, 8$) for which space-filling non-singular linear systems exist, and this will in fact be our principal objective in the Chapter. Since the case $n = 2$ has already been explained in detail (Ch. 3, § 2, Ex. 2, Ch. 4, § 1, Ex. 1 and § 3, Ex. 1), we begin with $n = 4$.

1. Clifford parallelism of solids in S_7

We are concerned in this section with Clifford parallelism of solids with respect to a 6-dimensional quadric Ω in S_7 (the case $n = 4$), and our principal objective is to find a simple geometrical construction for the important space-filling type of linear ∞^4-system (of Clifford parallel solids) that exists in this case. Of all the higher dimensional Clifford parallelisms in S_{2n-1}, the case $n = 4$ is easily the most interesting; and both here and in the following Chapters on representation theory we shall treat it in some detail.

Suppose then that X and Y are two skew solids which lie on $\Omega(=\Omega_6)$, and recall (Ch. 2, § 2) that Ω induces a natural correlation κ between X and Y. For convenience of notation we may regard κ as defining a mapping

$$\kappa: \tilde{X} \to \tilde{Y}$$

from the aggregate $\tilde{X}$ of lines of X to the aggregate $\tilde{Y}$ of lines of Y. We begin then by proving

LEMMA A. *Let l and m be two lines, lying in X and Y respectively, which span a solid Π. Then the polar solid Π' of Π (with respect to Ω) is that which spans the lines $l' = \kappa^{-1}(m)$ and $m' = \kappa(l)$.*

Proof. Since l and m' correspond by κ, the solid Σ which they span lies on Ω; and so, likewise, does the solid Σ' spanned by l' and m. Hence since l is the intersection of X with Σ, its polar [5] is the join of X to Σ, and therefore contains l' and m'; and, similarly, the polar [5] of m likewise contains l' and m'. Hence, since Π is spanned by l and m, its polar space Π' contains l' and m', and it is clearly spanned by them. This proves the Lemma.

We now select a non-singular null-polarity σ in X and keep it fixed throughout the following discussion.

Again, for notational convenience, we may regard σ as an (involutory) line–line mapping

$$\sigma: \tilde{X} \to \tilde{X}$$

in X. There is then, plainly, an associated null-polarity τ in Y given by

$$\tau = \kappa\sigma\kappa^{-1};$$

and if we put $\varpi = \tau\kappa$, so that also $\varpi = \kappa\sigma$, we see that ϖ (being a product of two correlations) is a collineation from (the lines of) X onto (the lines of) Y. The relations between κ, σ, τ and ϖ (in their capacity as line–line correspondences) may be summed up by saying that the diagram

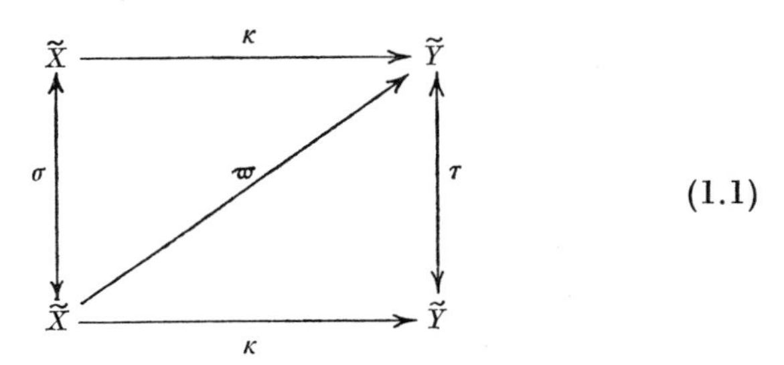

(1.1)

commutes. We now prove

Lemma B. *If two lines l and m, of X and Y respectively, correspond by the collineation ϖ, then so also do $l' = \sigma(l)$ and $m' = \tau(m)$. Also, the solid spanned by l' and m' is the polar of that spanned by l and m.*

Proof. The first assertion is that, if $m = \varpi(l)$, then

$$\tau(m) = \varpi(\sigma(l)),$$

i.e. that $\tau\varpi(l) = \varpi\sigma(l)$ for every l in X. This follows at once from (1.1).

As regards the second assertion, we know by Lemma A that the polar of the solid (l, m) is that spanned by $\kappa^{-1}(m)$ and $\kappa(l)$; so we have only to show that $\sigma(l) = \kappa^{-1}(m)$ and $\tau(m) = \kappa(l)$, i.e. that $\sigma(l) = \kappa^{-1}\varpi(l)$ and $\tau\varpi(l) = \kappa(l)$. These again follow immediately from (1.1), and the Lemma is proved.

Proposition 1.1. *If l_1 and l_2 are two lines of X such that l_1, l_2, $\sigma(l_1)$ and $\sigma(l_2)$ are four skew lines, and if m_1 and m_2 are their ϖ-images in Y, then the solid Π_1 spanned by l_1 and m_1 is Clifford parallel to the solid Π_2 spanned by l_2 and m_2.*

Proof. By Lemma B, the polar solid Π_1' of Π_1 is spanned by $\sigma(l_1)$ and $\tau(m_1)$, while the polar solid Π_2' of Π_2 is spanned by $\sigma(l_2)$ and $\tau(m_2)$. We have to show that Π_1, Π_2, Π_1', Π_2' belong to a regulus of solids.

Now two lines of S_3 and their images under a null-polarity always belong to a regulus (provided they are mutually skew).† So l_1, l_2, $\sigma(l_1)$ and $\sigma(l_2)$ belong to a regulus R_X of lines in X, and m_1, m_2, $\tau(m_1)$ and $\tau(m_2)$ belong to a regulus R_Y in Y; and the collineation ϖ maps R_X into R_Y in such a way that the lines of the former set are mapped respectively into the lines of the latter set. Hence, since X and Y are skew, it follows by Prop. 3.4 of Ch. 2 that the joins (l_1, m_1), (l_2, m_2), $(\sigma(l_1), \tau(m_1))$ and $(\sigma(l_2), \tau(m_2))$ belong to a regulus of solids in S_7, i.e. Π_1, Π_2, Π_1', Π_2' belong to a regulus R. This proves the Proposition.

The regulus R, as will now be apparent, is the Clifford regulus

† Almost all of what is proved in this section can be generalized to geometry in S_{2n-1}; but the particular assertion just made is the exception, being peculiar to S_3 (cf. Ex. 1 below).

containing Π_1 and Π_2, each generator of R being the join of a line of X to its ϖ-image in Y. Hence, from the definition of a linear system, we obtain at once the following result:

THEOREM 1.2. *Let X and Y be two skew solids on $\Omega = \Omega_6$; let κ be the natural correlation from X to Y induced by Ω; let σ be any non-singular null-polarity in X; and let ϖ denote the collineation $\kappa\sigma$ from X to Y. Then the system $\mathscr{F}$ of ∞^4 solids obtained by joining the ∞^4 lines of X to their respective ϖ-images in Y is a linear system of Clifford parallels.*

The system $\mathscr{F}$ contains in all ∞^6 Clifford reguli. Further, $\mathscr{F}$ contains ∞^3 solids which lie on Ω, namely those solids $(l, \varpi(l))$ for which l is self-polar with respect to σ.

We now observe, further, that the collineation ϖ between X and Y generates a regulus $\bar{R}$ (containing X and Y) whose directrices are the lines joining corresponding points of the two spaces. Also, since ϖ is of the form $\kappa\sigma$, it follows (cf. Ch. 3, § 6, Ex. 1) that $\bar{R}$ is an Ω-regulus. Each solid of $\mathscr{F}$ meets each generator of $\bar{R}$ in a line, i.e. it is a quadric-solid of $\bar{R}$; and since quadric-solids of $\bar{R}$ form an irreducible system of dimension 4, it follows that $\mathscr{F}$ is the system of all quadric-solids of the Ω-regulus $\bar{R}$. Hence:

THEOREM 1.3. *The solids of the system $\mathscr{F}$ defined in Theorem 1.2 are those and only those which meet every generator of a certain Ω-regulus $\bar{R}$ in a line, i.e. they are the quadric-solids of $\bar{R}$.*

We note:

(i) It is implicit in the above that $\mathscr{F}$ determines and is uniquely determined by the Ω-regulus $\bar{R}$. Now (cf. Ch. 3, § 6) the Ω-reguli (for $n = 4$) form two distinct ∞^{15}-families – reguli of α-solids and reguli of β-solids of Ω. It follows, then, that the linear systems of Clifford parallels projectively equivalent to $\mathscr{F}$ form two ∞^{15}-families.

(ii) From Prop. 3.3 of Ch. 2 it follows that *there is a unique solid of the system $\mathscr{F}$ through any point of S_7 that does not lie on $\bar{R}$.* Moreover since, by Th. 1.3, every solid of $\mathscr{F}$ contains ∞^1 mutually skew lines of Ω, it either meets Ω in a non-singular quadric surface or lies entirely on Ω, i.e. *each solid of* $\mathscr{F}$ *is either chordal or a generator of* Ω.

We remark finally that if Π_1, Π_2, Π_3 are any three generators of $\bar{R}$, then the quadric-solids of $\bar{R}$ are completely identified as all those solids which meet each of Π_1, Π_2, Π_3 in a line. Hence:

The system of all solids which meet three mutually skew generators of Ω *in lines is a linear system* $\mathscr{F}$ *of Clifford parallels,* i.e. *it is such a system* $\mathscr{F}$ *as has been described in Ths.* 1.2 *and* 1.3.

This gives a simple generation for such systems as $\mathscr{F}$ in terms of incidence properties only.

Ex. 1. Further to the footnote to Prop. 1.1, show that if n is even, if σ is a non-singular skew-polarity in a generator X of Ω, and if $\mathscr{G}$ is an irreducible ∞^r-system of $[\frac{1}{2}n-1]$'s in X which is (a) invariant under σ and (b) such that two general members of it together with their σ-images belong to a regulus, then σ and $\mathscr{G}$ may be used to construct, after the manner of Th. 1.2, an ∞^r-system of Clifford parallel $[n-1]$'s of S_{2n-1}. State a further condition on $\mathscr{G}$ which will ensure that this system of Clifford parallels is linear.

Ex. 2. If Σ is any generator of the regulus $\bar{R}$ mentioned in Th. 1.3, then the spaces of $\mathscr{F}$ are in unexceptional birational correspondence (by incidence) with the lines of Σ. Derive from this a mapping of the spaces of $\mathscr{F}$ on the points of a Klein quadric W_4^2 in S_5, and show that the pairs of polar spaces of $\mathscr{F}$ are represented on W_4^2 by pairs of points whose joins pass through a fixed point not on W_4^2. Obtain from this a mapping of the polar pairs of $\mathscr{F}$ on the points of a space S_4, and discuss the representation in this mapping of the Clifford reguli in $\mathscr{F}$.

2. The case $n = 4$ continued

Arising out of the work of the previous section, two questions remain to be answered.

(a) Is the system $\mathscr{F}$ which we constructed in § 1 a *non-singular* linear system?

(b) Are there any other types of linear system, non-singular or singular, in S_7?

To answer these questions we proceed directly from the equations developed in Ch. 4, the discussion for this case $n = 4$ being simplified by making use of the so-called *dual* matrices and their properties.

We remind the reader that if $S = (s_{ij})$ is a skew-symmetric matrix of order 4, then its dual S^D is defined to be the matrix $T = (t_{ij})$, also skew-symmetric, given by

$$t_{ij} = \frac{1}{2} \sum_{k=1}^{4} \sum_{l=1}^{4} \epsilon_{ijkl} s_{kl},$$

where ϵ_{ijkl} is the alternating symbol on four suffixes. If, in a different notation, we write

$$S = \begin{pmatrix} 0 & -\lambda_1 & -\lambda_2 & -\lambda_3 \\ \lambda_1 & 0 & -\lambda'_3 & \lambda'_2 \\ \lambda_2 & \lambda'_3 & 0 & -\lambda'_1 \\ \lambda_3 & -\lambda'_2 & \lambda'_1 & 0 \end{pmatrix},$$

we find that S^D is obtained from S by interchanging the primed with the unprimed letters. It follows at once that

$$(S^D)^D = S, \tag{2.1}$$

and we verify easily that

$$SS^D = S^D S = -(\lambda_1\lambda'_1 + \lambda_2\lambda'_2 + \lambda_3\lambda'_3)\, I_4 \tag{2.2}$$

and $$\det S = \det S^D = (\lambda_1\lambda'_1 + \lambda_2\lambda'_2 + \lambda_3\lambda'_3)^2.$$

From (2.2) we deduce that, if S is a non-zero skew-symmetric matrix of order 4 such that S^2 is a multiple of I_4, then S must be† a multiple of S^D, say $S = kS^D$. Moreover, taking duals of both sides of this equation, we obtain $S^D = kS$, so that $k^2 = 1$ and hence $k = \pm 1$. Hence there are two kinds of skew-symmetric matrix S of order 4 such that S^2 is a multiple of I_4:

(i) matrices S for which $S^D = S$, and

(ii) matrices S for which $S^D = -S$.

† Strictly speaking, this deduction only follows from (2.2) if S is non-singular; but it is easy to verify that the conclusion remains valid if S is singular.

Typical matrices of types (i) and (ii) are, respectively,

$$\begin{pmatrix} 0 & -\lambda_1 & -\lambda_2 & -\lambda_3 \\ \lambda_1 & 0 & -\lambda_3 & \lambda_2 \\ \lambda_2 & \lambda_3 & 0 & -\lambda_1 \\ \lambda_3 & -\lambda_2 & \lambda_1 & 0 \end{pmatrix} \text{ and } \begin{pmatrix} 0 & -\lambda_1 & -\lambda_2 & -\lambda_3 \\ \lambda_1 & 0 & \lambda_3 & -\lambda_2 \\ \lambda_2 & -\lambda_3 & 0 & \lambda_1 \\ \lambda_3 & \lambda_2 & -\lambda_1 & 0 \end{pmatrix}. \quad (2.3)$$

Moreover it is easy to check further that

(*a*) if S_1 and S_2 are both of type (i) or both of type (ii), then $S_1 S_2 + S_2 S_1$ is always a multiple of I_4;

(*b*) if S_1 and S_2 are of opposite types, then $S_1 S_2 + S_2 S_1$ is never a multiple of I_4 unless $S_1 = 0$ or $S_2 = 0$.

Hence, with reference to Ch. 4, Th. 3.1 and its proof, we deduce the following result:

If Ω has equation $x^T y = 0$ and R_0 is the Clifford regulus $y = \lambda x$, then any linear system of Clifford parallel solids in S_7 that contains R_0 is contained in one or other of the two ∞^4-systems

$$y = (\alpha I + S^{(1)}(\lambda))\, x, \quad y = (\alpha I + S^{(2)}(\lambda))\, x, \quad (2.4)$$

where $S^{(1)}(\lambda)$ and $S^{(2)}(\lambda)$ are the two matrices given in (2.3) *above.*

The two systems (2.4) are not, in fact, projectively distinct. Each is transformed into the other if we interchange x_1 with x_2 and y_1 with y_2 (with an interchange also of parameters λ_1 and λ_2), this being equivalent to a self-collineation of S_7 which leaves Ω invariant. It follows that either of the forms (2.4) is a canonical form to which the equations of *any* linear ∞^4-system of Clifford parallel solids in S_7 may be reduced. In particular the equations of the system $\mathscr{F}$ constructed in § 1 must be reducible to this form. We therefore have

THEOREM 2.1. *Every linear ∞^4-system of Clifford parallel solids in S_7 is projectively equivalent to the system $\mathscr{F}$ defined by the construction given in Theorems* 1.2 *and* 1.3.

Moreover, as we have remarked above, any system of Clifford parallel solids in S_7 is contained in a system projectively equivalent to $\mathscr{F}$ and this effectively answers question (*b*) which we posed at the beginning of this section. As regards question (*a*), the answer here is in the affirmative, as the reader may easily verify directly from equations (2.4) (cf. also Ex. 1 below). Hence, recapitulating, we have

COROLLARY 2.1.1. *Every linear system, non-singular or singular, of Clifford parallel solids in* S_7 *is contained in a non-singular linear* ∞^4*-system which is projectively equivalent to the system* $\mathscr{F}$ *defined in* § 1.

We might add, as we have already suggested in Ch. 4 (§§ 4, 5), that simple and concise results of the above kind are not to be expected in general for linear systems of Clifford parallels in S_{2n-1}.

Ex. 1. Using the notation defined in the Appendix, § 3, show that the matrices $S^{(1)}(\lambda)$ and $S^{(2)}(\lambda)$ in (2.4) can be expressed in the forms

$$S^{(1)}(\lambda) = \lambda_1 I \times J + \lambda_2 J \times K + \lambda_3 J \times L,$$

$$S^{(2)}(\lambda) = \lambda_1 K \times J + \lambda_2 J \times I + \lambda_3 L \times J,$$

and hence verify directly that the two systems given by (2.4) are non-singular.

Ex. 2. By the result given in Th. 1.3, there must exist Ω-reguli associated as in this Theorem with the two systems given by (2.4). Verify that these Ω-reguli are given, with $\epsilon = 1$ and $\epsilon = -1$ respectively, by the equations

$$x_0 + ix_1 = \lambda(x_2 + \epsilon i x_3),$$

$$x_0 - ix_1 = -\lambda^{-1}(x_2 - \epsilon i x_3),$$

$$y_0 + iy_1 = \lambda(y_2 + \epsilon i y_3),$$

$$y_0 - iy_1 = -\lambda^{-1}(y_2 - \epsilon i y_3).$$

Ex. 3. A Clifford regulus R in S_7 is contained in just two (non-singular) linear ∞^4-systems of Clifford parallel solids. The Ω-reguli with which these are associated (as in Th. 1.3) consist of solids of one and the same system on Ω. In fact, as we observed in Ch. 4, § 6, a notable feature of the case $n = 4$ is that the kernel of a Clifford regulus R, being the Segre product of a quadric surface with a line, is therefore the product of three lines; so that it may be regarded in *three separate ways* as the product of a quadric surface by a line. It lies, therefore, on three different reguli of solids of which one is the Clifford regulus R from which we started. The other two are the Ω-reguli associated with the two linear ∞^4-systems of Clifford parallels that contain R.

3. The Bridging Theorem

Having now discussed the case $n = 4$ in some detail, our next objective is to find geometrical constructions for systems of Clifford parallel [7]'s in S_{15} (the case $n = 8$), and most particularly for the third and last type of space-filling non-singular linear system – an ∞^8-system of [7]'s in S_{15}. In preparation for the subsequent attack on this problem, however, we first exhibit in the present section a general line of procedure for constructing linear systems of Clifford parallel $[n-1]$'s for any even value of n. This procedure as it happens, apart from its subsequent application to S_{15}, will also give us (cf. Ex. 5 below) a new construction for the space-filling linear systems of solids in S_7 – a construction that is no less simple, though quite different from that which we established in § 1.

Suppose then that n is even and that R is a Clifford regulus in S_{2n-1} and let K_R be the kernel of R. Further, let $\mathscr{F}$ be a linear system of Clifford parallels containing R. Then, as we have already noted (Ch. 4, § 4, Ex. 2), if $\mathscr{F}$ has weight $k > 1$, the generic space Π_0 of $\mathscr{F}$ meets R in two linear spaces $\bar{\Sigma}_0$ and $\bar{\Sigma}_0'$, each of dimension $\frac{1}{2}n-1$, which lie on K_R. *We now select for purposes of reference a fixed quadric generator q of K_R.* In the language of Ch. 3, Prop. 5.1, the two spaces $\bar{\Sigma}_0$ and $\bar{\Sigma}_0'$ then *lie over* a pair of generators Σ_0 and Σ_0' of q, and these can be shown to be skew to one another (cf. Th. 3.1 below). In this situation we shall say that the space Π_0 *bridges* the pair of generators (Σ_0, Σ_0') of q; it is, in fact, one of ∞^2 $[n-1]$'s of S_{2n-1} which bridge the pair (Σ_0, Σ_0'). The result given in Ch. 4, § 4, Ex. 2 may now be expressed by saying that the generic space Π_0 of a linear system $\mathscr{F}$, of weight $k > 1$, bridges† a non-intersecting pair of generators

† We emphasize that the role of q in this statement is only that of a selected base of reference, and any other quadric generator of K_R would serve the same purpose equally well. Further, the bridging phenomenon takes place relative to *any* quadric generator of *any* Clifford kernel within the system $\mathscr{F}$.

(Σ_0, Σ_0') of q. We propose now to improve this statement as follows.

THEOREM 3.1. *Let n be even; let $\mathscr{F}$ be a linear system of Clifford parallels with respect to Ω, containing the Clifford regulus R; and let q be any quadric generator of the kernel K_R of R. Then*

(i) *if the weight k of $\mathscr{F}$ is greater than* 1, *a generic space Π_0 of $\mathscr{F}$ bridges a non-intersecting pair (Σ_0, Σ_0') of generators of q;*

(ii) *if a second generic space Π_1 of $\mathscr{F}$ (independent of Π_0) bridges the pair (Σ_1, Σ_1'), then a sufficient condition for $\Sigma_0, \Sigma_0', \Sigma_1, \Sigma_1'$ to be four mutually skew spaces is $k > 2$ (a condition which is also necessary when $\mathscr{F}$ is maximal†); and*

(iii) *when this condition is satisfied, $\Sigma_0, \ldots, \Sigma_1'$ belong to a regulus of generators on q.*

Proof. By Th. 3.1 of Ch. 4 we may take the equation of Ω to be $x^T y = 0$, that of R to be $y = \lambda x$, and that of a generic space Π_0 of $\mathscr{F}$ to be

$$y = (\alpha I + \lambda_1 S_1 + \ldots + \lambda_{r-1} S_{r-1})\, x = \{\alpha I + S(\lambda)\}\, x, \qquad (3.1)$$

where $\alpha, \lambda_1, \ldots, \lambda_{r-1}$ are independent indeterminates and $S_1, \ldots, S_{r-1}$ satisfy the extended Hurwitz–Radon matrix equations (3.1) of Ch. 4. We may then choose q to be the terminal quadric of R given by

$$y = 0, \quad x^T x = 0.$$

We now find easily (cf. Ch. 4, §4, Ex. 2) that Π_0 bridges the pair of generators (Σ_0, Σ_0') of q with equations

$$\Sigma_0: \quad \{\Lambda I + S(\lambda)\}\, x = 0, \qquad (3.2)$$

$$\Sigma_0': \quad \{\Lambda I - S(\lambda)\}\, x = 0, \qquad (3.2)'$$

where $\Lambda^2 = -(\lambda_1^2 + \ldots + \lambda_{k-1}^2)$.‡ Since any common solution of (3.2) and (3.2)′ satisfies $\Lambda x = 0$, it follows that $\Lambda = 0$ is a necessary (and sufficient) condition for Σ_0 and Σ_0' to meet. But since $\alpha, \lambda_1, \ldots, \lambda_{r-1}$ are independent indeterminates, we see that

† The proof of this assertion of necessity will be deferred until later. See Ex. 4 at the end of this section and Ch. 6, §2, Ex. 1.

‡ Each of (3.2) and (3.2)′ represents n linear equations in $x_0, \ldots, x_{n-1}$; but these n equations reduce to $\frac{1}{2}n$ equations because each of the matrices $\Lambda I \pm S(\lambda)$ is of rank $\frac{1}{2}n$.

$\Lambda^2 = -(\lambda_1^2 + \ldots + \lambda_{k-1}^2)$ is zero if $k = 1$ but non-zero if $k > 1$. This proves (i) of the Theorem.

Now suppose that Π_1 has equation

$$y = (\alpha' I + \lambda_1' S_1 + \ldots + \lambda_{r-1}' S_{r-1}) x = \{\alpha' I + S(\lambda')\} x,$$

where $\alpha', \lambda_1', \ldots, \lambda_{r-1}'$ are further independent indeterminates, independent of $\alpha, \lambda_1, \ldots, \lambda_{r-1}$. Then Π_1 will bridge the pair (Σ_1, Σ_1') of generators of q given by

$$\Sigma_1\colon \quad \{\Lambda' I + S(\lambda')\} x = 0, \tag{3.3}$$

$$\Sigma_1'\colon \quad \{\Lambda' I - S(\lambda')\} x = 0, \tag{3.3$'$}$$

where $\Lambda'^2 = -(\lambda_1'^2 + \ldots + \lambda_{k-1}'^2)$. If Σ_0 were to meet Σ_1 (resp. Σ_1'), then (3.2) and (3.3) (resp. (3.3)$'$) would have a common solution for x; and then, if

$$T = \{\Lambda I + S(\lambda)\} + \{\Lambda' I \pm S(\lambda')\},$$

any such common solution would also satisfy $Tx = 0$, so that T would be singular. But if we write

$$\phi = \lambda_1 \lambda_1' + \ldots + \lambda_{k-1} \lambda_{k-1}',$$

we verify easily (using the relations (3.1) of Ch. 4, § 3) that

$$TT^T = 2(\Lambda\Lambda' \pm \phi) I;$$

so that T is singular if and only if $\Lambda\Lambda' \pm \phi = 0$. Since this equation is not satisfied (with either choice of sign) when $k > 2$ and the λ_i, λ_j' are independent indeterminates, it follows that Σ_0 does not meet Σ_1 or Σ_1' when $k > 2$; and a similar argument applies to Σ_0' and Σ_1 or Σ_1'. This proves (ii) of the Theorem.

Finally, to show that, when $k > 2$, the four spaces $\Sigma_0, \Sigma_0', \Sigma_1, \Sigma_1'$ belong to a regulus it is only necessary to prove that every line which meets Σ_0, Σ_0' and Σ_1 also meets Σ_1', i.e. if $x^{(1)}$ satisfies (3.2), if $x^{(2)}$ satisfies (3.2)$'$ and if $x^{(1)} + \alpha x^{(2)}$ satisfies (3.3), then there exists a scalar β such that $x^{(1)} + \beta x^{(2)}$ satisfies (3.3)$'$. The verification of this, amounting to an exercise in the properties of the extended Hurwitz–Radon matrix equations, may be left to the reader (cf. Ex. 1 below): and we note only that the appropriate value of β is given by

$$\beta = \frac{\phi - \Lambda\Lambda'}{\phi + \Lambda\Lambda'} \alpha.$$

This proves (iii), and the proof of Th. 3.1 is therefore complete.

We now investigate the converse situation and prove

THEOREM 3.2. (*The Bridging Theorem.*) *Let n be a multiple*† *of* 4 *and let q be any quadric-generator of the kernel K_R of a Clifford regulus R with respect to Ω. Then any space Π_0 which bridges a skew pair (Σ_0, Σ'_0) of generators of q is Clifford parallel to almost all spaces of R. Further, if a second space Π_1 bridges a skew pair (Σ_1, Σ'_1) of generators of q, and if Π_1 is skew both to Π_0 and to its polar space Π'_0, then Π_1 is Clifford parallel to Π_0 if Σ_0, Σ'_0, Σ_1, Σ'_1 belong to a regulus.*

(It will be noted that we have not precluded the possibility that Σ_1 may coincide with Σ_0 or Σ'_0, nor even, more specially, that the pair (Σ_1, Σ'_1) may coincide with the pair (Σ_0, Σ'_0).)

Proof. We shall prove the second of the two assertions in the theorem, observing that this includes the first as a special case because any generator of R bridges every pair of skew generators of q.

Suppose then that the spaces Σ_0, Σ'_0, Σ_1, Σ'_1 to which the Theorem refers belong to a regulus R_0. *We may assume then, without any loss of generality, that R_0 lies on q.* For if three at least of the spaces are distinct, then R_0 must certainly lie on q (since all its directrix lines must do so); and if, on the other hand, the pair (Σ_1, Σ'_1) coincides with the pair (Σ_0, Σ'_0), then we can take R_0 to be any one of the reguli on q containing Σ_0 and Σ'_0 (of which there are infinitely many since n is a multiple of 4).

We now choose a coordinate system (in S_{2n-1}) which, it may be noted, is not of either of the two standard kinds we have so far employed. First, since n is even, we may split up the $2n$ homogeneous coordinates into four column vectors ξ, ξ', η, η', each of length $\frac{1}{2}n$. We then choose two distinct generators, π and π' say, of the q-regulus R_0, and we arrange the coordinate system, as we evidently may, so that

(i) the ambient space of q is given by $\eta = \eta' = 0$;

(ii) the equation of q (in this ambient space) is $\xi^T\xi' = 0$; and

(iii) the equations of π and π' (in the same space) are $\xi = 0$ and $\xi' = 0$ respectively.

† For the case when n is an odd multiple of 2, see Theorem 3.2′ and the remarks preceding it.

Next, by an allowable coordinate transformation which does not affect (i), (ii) and (iii), we arrange further that

(iv) the generators of the Clifford regulus R are given, for variable μ, by the equations $\eta = \mu\xi$, $\eta' = \mu\xi'$.

We now recall (cf. Ch. 4, Cor. 6.1.1) that when a Clifford regulus and its kernel are given, the absolute quadric Ω is determined to the extent that it must belong to a certain net which, in the present coordinate system, is easily found to have the equation

$$a\xi^T\xi' + b(\xi^T\eta' + \eta^T\xi') + c\eta^T\eta' = 0; \tag{3.4}$$

and we may suppose, therefore, that (3.4) is the equation of Ω for certain fixed values of a, b and c such that $ac \neq b^2$ (the condition of non-singularity). With this equation for Ω, we find, then, that the polar of the generator of R with parameter μ is that with parameter $-(a+b\mu)/(b+c\mu)$; and the parameters of the terminal generators of R are the roots of the equation $c\mu^2 + 2b\mu + a = 0$.

Finally, we make the allowable transformation to new coordinates (x, x', y, y') given by

$$\left.\begin{aligned} x &= \alpha\xi + \beta\eta, & x' &= \alpha\xi' + \beta\eta' \\ y &= \gamma\xi + \delta\eta, & y' &= \gamma\xi' + \delta\eta' \end{aligned}\right\} \quad (\alpha\delta \neq \beta\gamma),$$

where α, β, γ, δ are chosen to satisfy

$$\begin{pmatrix} \alpha & \gamma \\ \beta & \delta \end{pmatrix}\begin{pmatrix} \alpha & \beta \\ \gamma & \delta \end{pmatrix} = \begin{pmatrix} a & b \\ b & c \end{pmatrix},$$

this choice being always possible because $ac \neq b^2$. The resulting new equation of Ω is then

$$x^Tx' + y^Ty' = 0, \tag{3.5}$$

and the new equations of R are

$$y = \lambda x, \quad y' = \lambda x',$$

the parameter λ being related to the parameter μ of (iv) above by the equation $\lambda = (\delta\mu + \gamma)/(\beta\mu + \alpha)$. Further, the generator of R which contains q is now given by the value $\gamma/\alpha = \theta$, say, of λ; the equation of q in this generator is

$$x^Tx' = 0;$$

and the equations of π and π' are $x = 0$ and $x' = 0$ respectively. The generators of the regulus R_0 on q will now be given (cf. Ch. 3, § 6) by an equation of the form

$$x' = \psi S x, \tag{3.6}$$

where S is a fixed non-singular skew-symmetric matrix of order $\frac{1}{2}n$ and ψ is a parameter. We may suppose, therefore, that the equations of $\Sigma_0, \Sigma_0', \Sigma_1, \Sigma_1'$ (inside the space $y = \theta x$, $y' = \theta x'$) are given by substituting $\psi_0, \psi_0', \psi_1, \psi_1'$ for ψ in (3.6).

The lengthy process of choosing a convenient coordinate system now pays its dividend, for the rest of the proof is quite straightforward. The space lying over Σ_0 in the generator of R given by $\lambda = \lambda_0$ plainly has equations

$$y = \lambda_0 x, \quad y' = \lambda_0 x', \quad x' = \psi_0 S x,$$

while that over Σ_0' in the generator given by $\lambda = \lambda_0'$ has equations

$$y = \lambda_0' x, \quad y' = \lambda_0' x', \quad x' = \psi_0' S x.$$

The bridging join Π_0 of these two spaces has equation

$$\begin{pmatrix} x' \\ y' \end{pmatrix} = \begin{pmatrix} \alpha_0 S & \beta_0 S \\ \gamma_0 S & \delta_0 S \end{pmatrix} \begin{pmatrix} x \\ y \end{pmatrix} \tag{3.7}$$

where
$$\left.\begin{aligned} \alpha_0 &= \frac{\lambda_0 \psi_0' - \lambda_0' \psi_0}{\lambda_0 - \lambda_0'}, \quad \beta_0 = \frac{\psi_0 - \psi_0'}{\lambda_0 - \lambda_0'}, \\ \gamma_0 &= -\lambda_0 \lambda_0' \frac{\psi_0 - \psi_0'}{\lambda_0 - \lambda_0'}, \quad \delta_0 = \frac{\lambda_0 \psi_0 - \lambda_0' \psi_0'}{\lambda_0 - \lambda_0'}. \end{aligned}\right\} \tag{3.8}$$

The polar Π_0' of Π_0 with respect to Ω (as given by (3.5)) then has equation

$$\begin{pmatrix} x' \\ y' \end{pmatrix} = \begin{pmatrix} \alpha_0 S & \gamma_0 S \\ \beta_0 S & \delta_0 S \end{pmatrix} \begin{pmatrix} x \\ y \end{pmatrix}. \tag{3.9}$$

Similarly, the equation of any space Π_1 bridging Σ_1 and Σ_1' is of the form

$$\begin{pmatrix} x' \\ y' \end{pmatrix} = \begin{pmatrix} \alpha_1 S & \beta_1 S \\ \gamma_1 S & \delta_1 S \end{pmatrix} \begin{pmatrix} x \\ y \end{pmatrix} \tag{3.10}$$

and its polar space Π_1' is given by

$$\begin{pmatrix} x' \\ y' \end{pmatrix} = \begin{pmatrix} \alpha_1 S & \gamma_1 S \\ \beta_1 S & \delta_1 S \end{pmatrix} \begin{pmatrix} x \\ y \end{pmatrix}. \tag{3.11}$$

Finally, since by hypothesis Π_0, Π'_0, Π_1, Π'_1 are mutually skew, a simple application of the matrix cross-ratio theorem (cf. also Ex. 2 below) confirms that the four spaces given by (3.7), (3.9), (3.10) and (3.11) belong to a regulus. Hence Π_1 is Clifford parallel to Π_0, and the proof of the Bridging Theorem is complete.

It will be noted that the parameter θ which determines the generator R containing q does not appear in any of the above equations; and this reflects the fact, already noted, that any other quadric-generator of K_R can be substituted for q. We may also draw attention to the relatively minor role of Ω in the Bridging Theorem. We see, in fact, that the Clifford parallelism is essentially determined by the Clifford kernel K_R of R, and that Ω plays thereafter no essential role. This agrees with what we have already said in Ch. 4, § 6.

More importantly, we note that if λ_0 and λ'_0 in (3.8) are replaced by $-1/\lambda'_0$ and $-1/\lambda_0$ respectively, then α_0 and δ_0 remain unaltered while β_0 and γ_0 are interchanged, i.e. the substitution converts (3.7) into (3.9), these being the equations representing Π_0 and Π'_0 respectively. Since the parameters λ, λ' of any polar pair of generators of R with respect to Ω are such that $\lambda\lambda' = -1$, we have the following result:

COROLLARY 3.2.1. *A space Π_0 and its polar space Π'_0 bridge the same pair (Σ_0, Σ'_0) of generators of q. More particularly, if Π_0 is the join of the spaces over Σ_0 and Σ'_0 that lie in generators T_1 and T_2 of R, and if T'_1 and T'_2 are the polar generators of T_1 and T_2 in R, then Π'_0 is the join of the spaces over Σ_0 and Σ'_0 that lie in T'_2 and T'_1 respectively.*

We next note the easily verifiable algebraic result:

COROLLARY 3.2.2. *If a 2×2 matrix*

$$A_0 = \begin{pmatrix} \alpha_0 & \beta_0 \\ \gamma_0 & \delta_0 \end{pmatrix}$$

can be parametrized in the form (3.8), then ψ_0 and ψ'_0 are the eigenvalues of A_0.

The spaces Π_0 and Π_1, given by (3.7) and (3.10), being Clifford parallel, must belong to a Clifford regulus R; and it is natural to ask which pairs of generators of q are bridged by the other spaces of R. To answer this we observe that the generic space of R has equation

$$\begin{pmatrix} x' \\ y' \end{pmatrix} = \begin{pmatrix} \alpha S & \beta S \\ \gamma S & \delta S \end{pmatrix} \begin{pmatrix} x \\ y \end{pmatrix}, \tag{3.12}$$

where

$$\begin{pmatrix} \alpha & \beta \\ \gamma & \delta \end{pmatrix} = A(\lambda)$$

is the 2×2 matrix, depending on the single parameter λ, defined in Ch. 2, § 4, Ex. 1 (Π_0 and Π_1 being given respectively by $\lambda = 0$ and $\lambda = \infty$). From the algebra of the Bridging Theorem it follows that the space (3.12) bridges two generators of the regulus R_0 given by $x' = \psi Sx$, the two appropriate values of ψ being, by Cor. 3.2.2 above, the eigenvalues of $A(\lambda)$. As λ varies, however (cf. Ch. 2, § 4, Ex. 2), the eigenvalues of $A(\lambda)$ are the roots of the equation

$$(\psi - \psi_0)(\psi - \psi_0') + \theta(\psi - \psi_1)(\psi - \psi_1') = 0,$$

where θ is a certain function of λ. If ψ_0, ψ_0', ψ_1 and ψ_1' are all distinct, the pairs of roots of this equation, as λ (and hence θ) varies, are pairs of mates in the unique involution determined by the pairs (ψ_0, ψ_0') and (ψ_1, ψ_1'). This gives

COROLLARY 3.2.3. *If Π_0 and Π_1 are two Clifford parallel $[n-1]$'s bridging respectively the pairs of generators (Σ_0, Σ_0') and (Σ_1, Σ_1') of a regulus R_0 on q (the spaces $\Sigma_0, \ldots, \Sigma_1'$ being distinct), then any other space Π_2 of the Clifford regulus joining Π_0 to Π_1 bridges a pair (Σ_2, Σ_2') of generators of R_0 belonging to the unique involution of generators of R_0 determined by the pairs (Σ_0, Σ_0') and (Σ_1, Σ_1').*

The modifications of this result relevant to the cases when $\Sigma_0, \ldots, \Sigma_1'$ are not all distinct may be left to the reader.

The Bridging Theorem, as we have stated it, applies only when n is a multiple of 4; but we may now point out that it still holds (though less significantly) if n is of the form $4m+2$. In fact, if $n = 4m+2$, then q is of dimension $n-2 = 4m$, and it cannot therefore contain any set of more than two mutually skew $[\frac{1}{2}n-1]$'s; so that, in the second part of the theorem, the pair (Σ_1, Σ_1') must coincide with (Σ_0, Σ_0'). The proof we gave does not apply in this case because q now contains no regulus such as R_0;

but the reader should have no difficulty in supplying an alternative proof to cover this case. We may thus assert

THEOREM 3.2′. *The Bridging Theorem is valid also (in so far as it is applicable) when n is of the form* $4m+2$.

As the final development in this section, designed to lead up to applications of the Bridging Theorem in the next section, we introduce the

DEFINITION. If n is a multiple of 4 and q is a non-singular $(n-2)$-dimensional quadric, then an irreducible system $\mathscr{Q}$ of (generally distinct) unordered pairs of generating $[\frac{1}{2}n-1]$'s on q will be called a *regular system* if it is such that two independent generic pairs of $\mathscr{Q}$ constitute four *distinct* spaces of a regulus. (This implies that $\mathscr{Q}$ must be of positive dimension.) Further, if $\mathscr{Q}$ has the additional property that it is not contained in any more ample regular system on q, then it will be called a *maximal regular system* of pairs of generators of q.

In terms of the above definition the role of Theorems 3.1 and 3.2 in relation to the general problem of constructing linear systems of Clifford parallels can be described as follows.

Let R be a Clifford regulus in S_{2n-1}, where n is a multiple of 4, and let q be a quadric generator of the kernel K_R of R. Further let $\mathscr{Q}$ be any $(r-2)$-dimensional regular system of pairs of generating $[\frac{1}{2}n-1]$'s of q. Then $\mathscr{Q}$ gives rise by the bridging construction to an r-dimensional irreducible system $\mathscr{F}$ of Clifford parallel $[n-1]$'s of the space S_{2n-1} containing R, of which ∞^2 arise from each pair of $\mathscr{Q}$. The system $\mathscr{F}$ so arising will not always be linear; but we now assert that if $\mathscr{Q}$ is a *maximal* regular system, then $\mathscr{F}$ is linear and maximal. For if not, then $\mathscr{F}$ would be contained (by the Embedding Theorem, Ch. 4, Th. 2.1) in a strictly larger maximal linear system $\overline{\mathscr{F}}$; and (by Th. 3.1) the spaces of $\overline{\mathscr{F}}$ would bridge the pairs of a regular system $\overline{\mathscr{Q}}$ which strictly contains $\mathscr{Q}$ – a contradiction of the assumed maximality of $\mathscr{Q}$. Moreover, if we grant the validity of the necessity clause inserted (in parentheses) in Th. 3.1 (ii), then we can add that the derived system $\mathscr{F}$, in addition to being linear and maximal, is necessarily of weight $k>2$.

Conversely, if $\mathscr{F}$ is a maximal linear system, of weight $k > 2$, which contains R, then by Th. 3.1 the spaces of $\mathscr{F}$ bridge the pairs of a regular system $\mathscr{Q}$ on q. We assert then that this system $\mathscr{Q}$ is maximal. For if there existed a more ample regular system $\overline{\mathscr{Q}}$ containing $\mathscr{Q}$, then the spaces bridging pairs of $\overline{\mathscr{Q}}$ would form an irreducible system $\overline{\mathscr{F}}$ of Clifford parallels which (i) strictly contains $\mathscr{F}$ and (ii) is contained (by the Embedding Theorem) in a linear system $\overline{\mathscr{F}}'$; and this system $\overline{\mathscr{F}}'$ would then be a linear system strictly containing $\mathscr{F}$, in contradiction to the maximality of $\mathscr{F}$.

Recapitulating then we have

THEOREM 3.3. *When n is a multiple of 4, the existence of maximal $(r-2)$-dimensional regular systems of pairs of $[\frac{1}{2}n-1]$'s on an $(n-2)$-dimensional quadric q is equivalent to the existence of r-dimensional maximal linear systems of Clifford parallel $[n-1]$'s of S_{2n-1} of weight $k > 2$.*

For some remarks on the bridging properties of linear systems of weight 2 we refer the reader to Exs. 6 and 7 below. Systems of weight 1 (by Th. 3.1 (i)) have no bridging properties.

Except for the case $n = 8$, to be discussed in the next section, we shall not pursue the study of the bridging construction any further in this book. An obvious outstanding problem, however, is that of singling out some property of a maximal regular system which will ensure that the derived linear system of Clifford parallels is non-singular.

Ex. 1. Complete the proof of the last part of Th. 3.1.

[Hint: It was supposed that

$$\{\Lambda I + S(\lambda)\} x^{(1)} = 0, \quad \{\Lambda I - S(\lambda)\} x^{(2)} = 0 \tag{1}$$

and

$$\{\Lambda' I + S(\lambda')\} (x^{(1)} + \alpha x^{(2)}) = 0. \tag{2}$$

Premultiply (2) by $S(\lambda)$ and use (1) and the extended Hurwitz–Radon matrix equations to prove that

$$\{\Lambda' I - S(\lambda')\} (x^{(1)} - \alpha x^{(2)}) = -2\phi\Lambda^{-1}(x^{(1)} + \alpha x^{(2)}). \tag{3}$$

Now solve (2) and (3) to obtain $S(\lambda')\, x^{(1)}$ and $S(\lambda')\, x^{(2)}$ and hence prove that

$$\{\Lambda' I - S(\lambda')\} (x^{(1)} + \beta x^{(2)}) = 0,$$

where $\beta = \alpha(\phi - \Lambda\Lambda')/(\phi + \Lambda\Lambda')$.]

Ex. 2. Verify that the spaces given by equations (3.7), (3.9), (3.10) and (3.11) do in fact belong to a regulus. [Hint: If A_0 and A_1 denote the respective matrices

$$\begin{pmatrix} \alpha_0 & \beta_0 \\ \gamma_0 & \delta_0 \end{pmatrix} \quad \text{and} \quad \begin{pmatrix} \alpha_1 & \beta_1 \\ \gamma_1 & \delta_1 \end{pmatrix},$$

then the matrices appearing in these equations may be written in the forms

$$A_0 \times S, \quad A_0^T \times S, \quad A_1 \times S, \quad A_1^T \times S,$$

where $\times$ denotes Kronecker product of matrices. The required result now follows from Ch. 2, § 4, Ex. 1, using the basic multiplication rule $(A \times B)(C \times D) = AC \times BD$.]

Ex. 3. State a property of a regular system $\mathscr{Q}$ which will ensure that the derived bridging system of Clifford parallels is linear (cf. Cor. 3.2.3).

Ex. 4. If Σ_0, Σ_0', Σ_1, Σ_1' are four distinct generators of a regulus R_0 on q, show that the *smallest* linear system of Clifford parallels which contains all spaces bridging (Σ_0, Σ_0') and all spaces bridging (Σ_1, Σ_1') must also contain all spaces bridging (Σ_2, Σ_2'), where (Σ_2, Σ_2') is any pair of the unique involution τ of generators of R_0 determined by the pairs (Σ_0, Σ_0') and (Σ_1, Σ_1'). [Hint: use Cor. 3.2.3.]

It is in fact the case (as will emerge in Ch. 6, § 2, Ex. 1) that the ∞^3-system of spaces which bridge some pair of τ is linear and non-singular – and so of weight $k = 3$. Assuming this, prove the assertion as to necessity in Th. 3.1 (ii).

Ex. 5. *Bridging construction for* $n = 4$. When $n = 4$, the bridging construction for a 4-dimensional linear system of Clifford parallel solids in S_7 must be based on a 2-dimensional regular system of pairs of generating lines of a quadric surface q. In fact the system of *all* pairs of generators of one system on q is such a regular system; and the derived construction, quite different from that projectively given in § 1, may be stated as follows:

If q is any quadric-generator of the kernel of a Clifford regulus in S_7, then all the solids of S_7 that bridge pairs of generating lines of one system on q form a (maximal) linear ∞^4-system of Clifford parallels in S_7.

Ex. 6. When n is of the form $4m+2$, a quadric q_{n-2} cannot contain more than two mutually skew generators and so, in particular, cannot carry any regular system of pairs of generators. Also, by Th. 3.1, there can exist no linear systems of Clifford parallels of weight $k > 2$ in this case (a result in conformity with Th. A 1 of the Appendix). However, bearing in mind that we did not assume, in the Bridging Theorem, that Σ_0, Σ_0', Σ_1, Σ_1' were all distinct, it is still possible to obtain certain linear systems of weight two by a bridging construction. Namely, we consider the system of ∞^2 spaces which bridge a *fixed* pair of skew generators of q and this, as the reader may show, is linear and non-singular (cf. also Ch. 6, § 2, Ex. 1). When $n = 2$, this construction reduces to the standard one for this case given in Ch. 3, § 2, Ex. 2.

Ex. 7. An example of a linear system of weight two (when n is a multiple of 4), necessarily *not* obtainable by bridging a regular system, is the *half-fixed system* defined as follows:

Consider a system of pairs (Σ_0, Σ') of generators of q_{n-2}, in which Σ_0 is a fixed generator and Σ' varies arbitrarily in the complete system of generators of q to which Σ_0 belongs (a system of freedom $\frac{1}{8}n(n-2)$). Although these pairs do not form a regular system, the Bridging Theorem shows that the spaces bridging all such pairs form an irreducible system of Clifford parallels of freedom $\frac{1}{8}n(n-2)+2$. The system so obtained is in fact linear of weight two. (The weight cannot be greater than two by Th. 3.3, nor less than two by the proof of Th. 3.1.) The system is, indeed, projectively equivalent to that discussed in Ch. 4, § 4, Ex. 5.

4. Clifford parallelism of [7]'s in S_{15}

By Th. 3.3 of the preceding section with $n = 8$, we know that the problem of constructing maximal linear systems, of weight $k > 2$, of Clifford parallel [7]'s in S_{15} is equivalent to that of finding maximal regular systems of pairs of solids on a 6-dimensional (non-singular) quadric q_6. We now propose to identify all possible systems of this latter kind.

The first of two preliminary results that we shall need is

PROPOSITION† 4.1. *On the Grassmannian $G(3,7)$ of solids of S_7, the generating solids of either system on a non-singular quadric q_6 are represented by the points of a variety V_6^{128} which is the quadratic Veronesean of another non-singular quadric q_6^*.*

Proof. We choose a coordinate system in which q_6 has equation

$$x^T y \equiv x_0 y_0 + x_1 y_1 + x_2 y_2 + x_3 y_3 = 0$$

and observe that the generic solid of one system on q_6 has equation

$$y = Sx, \tag{4.1}$$

where $S = (s_{ij})$ is a skew-symmetric matrix of order 4 whose elements above the main diagonal are independent indeterminates $s_{12}, \ldots, s_{34}$. It is easy then to check that all the Grassmann coordinates of the solid (4.1), i.e. all determinants of order 4 extracted from the 4×8 matrix $(S, -I)$, can be written as quadratic forms in $z_0, \ldots, z_7$, where

$$z_0 = 1, \quad z_1 = s_{12}, \quad z_2 = s_{13}, \quad z_3 = s_{14},$$
$$z_4 = s_{23}, \quad z_5 = s_{24}, \quad z_6 = s_{34},$$
$$z_7 = s_{12}s_{34} - s_{13}s_{24} + s_{14}s_{23};$$

and, conversely, that every quadratic form in $z_0, \ldots, z_7$ is expressible as a linear combination of the Grassmann coordinates of (4.1). This shows that the Grassmannian of the system of solids with (4.1) as generic member is the quadratic Veronesean of the quadric q_6^* which is given, in the space of the coordinates $z_0, \ldots, z_7$, by the equation

$$z_0 z_7 = z_1 z_6 - z_2 z_5 + z_3 z_4;$$

and this proves the Proposition.

The second result that we need is

PROPOSITION 4.2. *On the Grassmannian $G(3,7)$ any regulus of solids of S_7 is represented by a rational normal quartic curve.*

The proof of this is elementary and may be left to the reader. (For a generalization, see Ch. 7, Prop. 1.2.)

† This proposition forms the basis of the so-called Study triality, of which we shall have more to say in Ch. 7.

Consider now the mapping of the solids of one system on q_6 – say the α-solids – by the points of q_6^*. A regulus of these solids, being represented on $G(3, 7)$ by a curve ${}^0C^4$ (Prop. 4.2), will be represented on q_6^* by a curve whose quadratic Veronesean is a ${}^0C^4$, i.e. by a conic. Hence:

Four α-solids of q_6 belong to a regulus if and only if their image points on q_6^* lie on a (proper) conic, section of q_6^* by a plane.

From this we deduce that a regular system of pairs of α-solids on q_6 must be represented on q_6^* by a system of point-pairs such that two independent generic pairs of the system are four distinct points of a conic. This implies, in particular, that the chords joining pairs of points of the system on q_6^* must all intersect one another. Thus the problem of finding regular systems of pairs of α-solids on q_6 is reduced essentially to that of finding systems of lines (in the ambient [7] of q_6^*) of which any two intersect. Such a system of lines, however, must be either a star of lines or a ruled plane (or a sub-system of one or other of these). In the former case the vertex of the star must not lie on q_6^*; and in the latter case the plane must cut q_6^* in a proper conic.† We have now proved

PROPOSITION 4.3. *Every maximal regular system of pairs of α-solids on q_6 belongs to one or other of two types:* (i) *the type represented on q_6^* by pairs of points whose joins all pass through a fixed point O not on q_6^*, and* (ii) *the type represented on q_6^* by all the point-pairs of the section of q_6^* by a non-tangent plane. Systems of these two types are of dimension* 6 *and* 2 *respectively.*

A maximal regular system of the first type will be called a *Study involution* of pairs of α-generators of q_6.

From Prop. 4.3 we deduce, by applying Th. 3.3, that there exist exactly two distinct types of maximal linear system, of weight $k > 2$, of Clifford parallel [7]'s in S_{15}, namely one type of dimen-

† If, in the former case, the vertex of the star were on q_6^*, or if, in the latter case, the plane met q_6^* in a reducible conic, then the corresponding system of solids on q_6 would not be regular according to our definition. A case can be made for extending the definition of a regular system to include such specialized systems, but we shall not pursue this idea here.

sion 8 and one of dimension 4. As regards the first type – the most important since it is space-filling and non-singular (cf. Ex. 1 below) – we have

THEOREM 4.4. *If τ is a Study involution of pairs of generating solids of one system on any quadric-generator of the kernel of a Clifford regulus in S_{15}, then the* [7]*'s which bridge pairs of τ form a maximal linear (space-filling) system of Clifford parallels in S_{15}.*

This system is, in fact, the complex analogue of the well-known system, in real elliptic 15-dimensional space, which gives rise to the fibering of the 15-sphere by 7-spheres (cf. Ch. 1, § 5).

In the examples below we give algebraic details of each of the two types of maximal system that we have described above.

EX. 1. *Systems of the first type.* If, in the ambient space of q_6^* (in which we took current coordinates $z_0, \ldots, z_7$) we take O to be the point $(1, 0, 0, 0, 0, 0, 0, -1)$ – not on q_6^* – we find that the associated maximal regular system of pairs of solids of q_6 has generic pair

$$x' = Sx, \quad x' = -(S^D)^{-1}x,$$

where S is a generic skew-symmetric matrix of order 4 (and S^D denotes the dual matrix, introduced in § 2). Then, using the Bridging Theorem, we may derive the equations of the associated linear ∞^8-system of Clifford parallels (after a minor notational change) in the form

$$\left.\begin{aligned} y &= Ax' + a'x, \\ y' &= A^D x + ax', \end{aligned}\right\}$$

where A is again skew-symmetric and a, a' are scalars (so that there are 8 parameters in all, as expected).

This, of course, is the form of the equations relative to Ω in the form $x^T x' + y^T y' = 0$. It is a straightforward matter to convert these equations to the canonical form employed in Ch. 4 (with Ω as $x^T y = 0$) and we find then that the system becomes

$$y = (\alpha I + \lambda_1 S_1'' + \ldots + \lambda_7 S_7'')\, x,$$

where $(S_1'', \ldots, S_7'')$ is the solution of the Hurwitz–Radon matrix equations given in the Appendix, § 3 (iii). The system is consequently non-singular and (by Ch. 4, Cor. 4.2.1) space-filling of index 1.

Ex. 2. If, in the ambient space of q_6^*, we draw rays through a fixed point O lying *on* q_6^*, we thereby generate the system of pairs (Σ_0, Σ') of generators of q_6 in which Σ_0 (the generator represented by O) is fixed, and Σ' varies in the complete system of generators of q_6 to which Σ_0 belongs. By bridging this system of pairs, we obtain the half-fixed system (for $n = 8$) described in § 3, Ex. 7 and, in this sense at least, the half-fixed system may be regarded as a limiting form of the non-singular system described in Ex. 1.

Ex. 3. *Systems of the second type.* The section of q_6^* by a (non-tangent) plane represents, as we saw earlier, a regulus of generators on q_6, so that the maximal regular systems of the second type are constituted by all the (unordered) pairs of generators of a fixed regulus on q_6. Thus we may take a generic pair of the system to be given by

$$x' = \psi_0 Sx, \quad x' = \psi_0' Sx,$$

where S is now a fixed skew-symmetric matrix of order 4 and ψ_0, ψ_0' are independent indeterminates. The equation of the associated bridging system is then precisely equation (3.7) of § 3 – with $\alpha_0, \beta_0, \gamma_0, \delta_0$ regarded as the variable parameters.

If the system is transformed to the canonical form of Ch. 4 (with Ω as $x^T y = 0$) we find that it reduces to

$$y = (\alpha I + \lambda_1 S_1 + \lambda_2 S_2 + \lambda_3 S_3)\, x,$$

where (S_1, S_2, S_3) is an 'extra' solution of the Hurwitz–Radon equations for $n = 8, r = 4$, as given in the Appendix, § 3 (iii). The system is accordingly non-singular (and maximal) but not, of course, space-filling. (Cf. also Ch. 6, § 2, Ex. 1.)

Ex. 4. Any two maximal systems of the same type are projectively equivalent (by an Ω-preserving collineation in S_{15}); the two types, in fact, belong to aggregates of respective total freedoms 84 and 100.

CHAPTER 6

THE T-REPRESENTATION

In the theory of Clifford parallel spaces, as so far developed, the reader will have noted that the essential unit is not the individual $(n-1)$-dimensional space Π of S_{2n-1} but the pair of polar spaces (Π, Π') with respect to the absolute quadric Ω. Parallelism, in effect, is a relation between two such polar pairs; and Clifford reguli, as well as linear systems of Clifford parallels in general, are all systems of polar pairs. What we now propose to do in the present Chapter is to set up a birational mapping of the polar pairs (Π, Π') of S_{2n-1} on the points of a certain algebraic variety T, of dimension n^2; and this mapping – the T-representation – will be such that the polar pairs belonging to any Clifford regulus in S_{2n-1} are mapped on T by the points of a line. It will then follow that the polar pairs of any linear system of Clifford parallels in S_{2n-1} are mapped on T by the points of a flat space. The T-representation which we define in this way is different from other representations of the polar pairs (Π, Π') of S_{2n-1} – to be discussed in Ch. 7 – which arise from $(2, 1)$-projections of the Grassmannian $G(n-1, 2n-1)$.

1. Basis of the representation

The idea of the T-representation is to associate with any general pair (Π, Π') of polar $[n-1]$'s with respect to Ω a unique quadric primal $\mathscr{Q}(\Pi, \Pi')$, and then to take the coefficients in the equation of $\mathscr{Q}(\Pi, \Pi')$ as the coordinates of the pair.

More precisely, supposing that Π and Π' are chordal, we denote by q, q' the $(n-2)$-dimensional quadrics in which Π and Π' meet Ω, and we consider the quadric $[n-1]$-cones $\Pi(q')$ and $\Pi'(q)$ that join Π to q' and Π' to q. Each of these cones meets Ω in the quartic $(2n-3)$-dimensional variety generated by all the lines joining points of q to points of q', so that the pencil of quadrics defined by $\Pi(q')$ and $\Pi'(q)$ contains Ω. Within this pencil, then, there is a unique quadric $\mathscr{Q}(\Pi, \Pi')$ which is the harmonic conjugate of Ω with respect to $\Pi(q')$ and $\Pi'(q)$, and this is the quadric that we propose to associate with the pair (Π, Π'). The birational character of the correspondence between polar pairs (Π, Π') and quadrics $\mathscr{Q}(\Pi, \Pi')$ follows at once from the observation that the pencil defined by $\Pi(q')$ and $\Pi'(q)$ contains no other cones.

To identify $\mathscr{Q}(\Pi, \Pi')$ algebraically, we choose the coordinate system so that Ω has equation $x^T y = 0$; and we suppose then, as we may, that Π and Π' have equations $y = Ax$ and $y = -A^T x$, where A is a matrix of order n. A straightforward calculation then gives the equations of the cones $\Pi(q')$ and $\Pi'(q)$ in the respective forms

$$K_1 \equiv (y - Ax)^T (A + A^T)^{-1} (y - Ax) = 0,$$

$$K_2 \equiv (y + A^T x)^T (A + A^T)^{-1} (y + A^T x) = 0,$$

provided that $A + A^T$ is non-singular, i.e. that Π is skew to Π'. We now write $(A + A^T)^{-1} = B = B^T$, so that

$$K_1 \equiv y^T B y - x^T A^T B y - y^T B A x + x^T A^T B A x,$$

$$K_2 \equiv y^T B y + x^T A B y + y^T B A^T x + x^T A B A^T x.$$

By the definition of B

$$BA + BA^T = AB + A^T B = I,$$

i.e. $$BA^T = I - BA \quad \text{and} \quad A^T B = I - AB$$

so that $$ABA^T = A^T BA \quad (= A - ABA).$$

Hence

$$K_2 - K_1 \equiv x^T (A + A^T) By + y^T B (A + A^T) x \equiv 2x^T y,$$

and this exhibits Ω as a member of the pencil given by

$$K_2 + \lambda K_1 = 0.$$

Now $\mathscr{Q}(\Pi, \Pi')$ was defined to be the harmonic conjugate of Ω with respect to the two cones $K_1 = 0$, $K_2 = 0$ of the pencil, so that its equation is $K_2 + K_1 = 0$; and this, after some reduction, gives the equation of $\mathscr{Q}(\Pi, \Pi')$ in the form

$$x^T A(A+A^T)^{-1} A^T x + x^T(A-A^T)(A+A^T)^{-1} y + y^T(A+A^T)^{-1} y = 0. \quad (1.1)$$

We take the coefficients in this equation, i.e. effectively the elements of the matrix triad

$$[A(A+A^T)^{-1} A^T, (A-A^T)(A+A^T)^{-1}, (A+A^T)^{-1}], \quad (1.2)$$

to be the *coordinates of the polar pair* (Π, Π')—the coordinates, in other words, of the image point of $\mathscr{Q}(\Pi, \Pi')$ in the space S_N $(N = \frac{1}{2}(2n-1)(2n+2))$ of quadrics of S_{2n-1}. If the elements a_{ij} of A are taken to be independent indeterminates, then the elements of (1.2) are the coordinates of the generic point of an *image variety* T of the polar pairs (Π, Π'). It should be noted, however, that although (1.2) exhibits T as a rational transform of the (affine) space S_{n^2} of the a_{ij}, the associated correspondence between T and S_{n^2} is one–two, since the same point of T arises equally from A and $-A^T$. We shall also have to consider the fundamental elements in the mapping of the pairs (Π, Π') by points of T, such as arise when Π and Π' are not skew to one another, so that $A + A^T$ is singular.

The variety T, as above defined, lies in a prime S_{N-1} of S_N. For the pencil of quadrics $K_2 + \lambda K_1 = 0$ has the property that any two of its members which separate the base cones $K_1 = 0$ and $K_2 = 0$ harmonically are mutually apolar; and indeed each of them is self-reciprocal with respect to the other.† Thus $\mathscr{Q}(\Pi, \Pi')$ is always apolar to Ω, so that its image point in S_N lies in a fixed prime S_{N-1} (cf. Ex. 1 below).

Note. There are other ways of constructing the variety T. Thus, for example, if $\mathscr{T}$ is the image variety of all the quadric cones $\Pi(q')$, i.e. all the tangent cones to Ω from spaces Π, and if O is the image point of Ω in S_N, then T is the (2,1) projection of $\mathscr{T}$ from O onto a prime S_{N-1}. Alternatively we may define T

† This follows from the observation that a coordinate system can be chosen in S_{2n-1}, with Π, Π' as reference spaces X, Y, such that the cones and any pair of quadrics separating them harmonically have equations $x^T x = 0$, $y^T y = 0$ and $x^T x \pm y^T y = 0$.

as the model of all the *pencils* of quadrics such as $K_2+\lambda K_1 = 0$. See also Ex. 3 below.

We now prove the following basic property of the T-representation.

THEOREM 1.1. *On the T-model as above defined, every Clifford regulus of S_{2n-1} is represented on T by a line;*† *and every linear system of Clifford parallels is represented on T by a linear space.*

Proof. Let R_0 be any Clifford regulus of S_{2n-1} and let the coordinate system be chosen so that Ω has equation $x^Ty = 0$ and R_0 is given by $y = \lambda x$. Then the equations of any pair (Π, Π') of polar spaces of R_0 are of the form $y = \pm\lambda x$ $(\lambda \neq 0)$. On making the substitution $A = \lambda I$ in (1.1) we obtain the equation of $\mathscr{Q}(\Pi, \Pi')$ in the form

$$\lambda^2 x^T x + y^T y = 0;$$

and, since the coefficients in this equation are linear in the single parameter λ^2, it follows that R_0 is represented on T by a line. Further, since our definition of the T-model was independent of any choice of the coordinate system in S_{2n-1}, it follows that *every* Clifford regulus is represented on T by a line.

To see now, further, that every linear system $\mathscr{F}$ of Clifford parallels in S_{2n-1} is represented on T by a linear space, it is only necessary to point out that $\mathscr{F}$ must be represented on T by an irreducible variety $\mathscr{V}$ with the property that $\mathscr{V}$ contains the line joining a generic pair of points of itself, i.e. $\mathscr{V}$ is a flat space. This completes the proof.

Ex. 1. Show that the linear relation (apolarity to Ω) satisfied by all the quadrics $\mathscr{Q}(\Pi, \Pi')$ is reflected in the fact that the matrix $(A-A^T)(A+A^T)^{-1}$ has zero trace.

Ex. 2. Show that the 'dual' T-model, obtained by assigning to (Π, Π') the coefficients in the tangential equation of $\mathscr{Q}(\Pi, \Pi')$, is projectively equivalent to the T-model. Deduce that the quadrics associated with the polar-pairs (Π, Π') of a linear system of Clifford parallels form a linear system of quadrics which is also a linear system of quadric envelopes.

† This result explains why it is natural to say that a Clifford regulus is the 'join' of any two of its generators.

Ex. 3. If Π and Π' (as polar $[n-1]$'s with respect to Ω) have equations $y = Ax$ and $y = -A^T x$, show that the harmonic biaxial involution with Π and Π' as axes is given by

$$\begin{pmatrix} x \\ y \end{pmatrix} \to \begin{pmatrix} (A+A^T)^{-1}(A^T-A) & 2(A+A^T)^{-1} \\ 2A(A+A^T)^{-1}A^T & (A-A^T)(A+A^T)^{-1} \end{pmatrix} \begin{pmatrix} x \\ y \end{pmatrix}$$

and deduce from this an alternative method of defining the T-model.

Ex. 4. *The T-models for $n = 2$ and 3.* We give here some preliminary details about these two models in advance of their further consideration in § 3.

(i) For $n = 2$, we find easily, from the parametric equations (1.2), that the T-model is a $V_4^6[8]$, Segre product of two planes, lying in a prime of the image space S_9 of quadrics of S_3. The two kinds of linear ∞^2-systems of Clifford parallel lines of S_3 are represented on T by the planes of its two generating systems.

(ii) For $n = 3$, we may identify the T-model as follows. If Ω in S_5 has equation $x_0 y_0 + x_1 y_1 + x_2 y_2 = 0$, then the coordinates $(x_0, x_1, x_2, y_0, y_1, y_2)$ of a point of Ω can be regarded as those of a line of S_3; and a pair of polar planes (Π, Π') with respect to Ω meet Ω in conics q, q' which represent complementary reguli on a quadric ψ of S_3. Further, the complex of tangent lines to ψ (the quadric complex determined by ψ) corresponds to the quartic threefold on Ω generated by all the lines meeting q and q'; and this threefold is the base of a pencil of quadrics which contains Ω. Then $\mathcal{Q}(\Pi, \Pi')$, being the unique quadric of this pencil which is apolar to Ω, is seen to be that given (essentially) by the *standard* equation of the quadric complex determined by ψ, i.e. that in which the coefficients are all the minors of order two from the matrix of ψ. The generic point of the resulting T-model is accordingly that whose coordinates in S_{19} are all the minors of order two extracted from a symmetric matrix of order four with independent indeterminates as elements; and, from this form of its generic point, the T-model is immediately identified as the known† variety $V_9^{92}[19]$.

† This variety has been discussed in detail by Todd[**26**] in connection with his work on the Cayley model of conics in S_3, and further properties of it were given by Semple[**21**] in his investigations of complete quadrics.

2. The T-representation of augmented generators

As we have already remarked, the formulae (1.2) which give the coordinates of the image point of a polar pair (Π, Π') no longer apply when Π meets Π', the matrix $A + A^T$ being then singular. We shall return later to the general problems arising from this; but for the moment we are concerned with the extreme case in which Π' coincides with Π, i.e. with a generator α or β of Ω which counts as a coincident polar pair (α, α) or (β, β). What we now wish to show is that whereas such a coincident pair, say (α, α), is fundamental for the transformation, being such that its total image in T is a linear space A of dimension $\frac{1}{2}(n-1)(n+2)$, each *augmented generator* $\alpha(q)$ based on α (cf. Ch. 3, § 5) corresponds in a well-defined manner to its own particular point of A. Each $\alpha(q)$, in fact, where q is any non-singular quadric in α, functions as a specialized limit of a polar pair (Π, Π') as Π and Π' both approach α within a Clifford regulus which has α as one of its terminal generators and q as the corresponding terminal quadric; and to this specialized limit there corresponds a definite point of A. More precisely we prove

PROPOSITION 2.1. *In the birational mapping of the polar pairs* (Π, Π') *of* S_{2n-1} *on the points of* T, *each generator* α *or* β *of* Ω *(regarded as a coincident polar pair) is fundamental, having a space* A *or* B, *of dimension* $\frac{1}{2}(n-1)(n+2)$, *as its total image on* T. *Any general point of* A *corresponds in a well-defined way to a unique augmented generator* $\alpha(q)$; *and similarly for* B. *As* α *varies, the image points of all the augmented generators of the type* $\alpha(q)$ *lie on a primal* $\mathscr{A}$ *of* T, *the* (n^2-1)*-dimensional subvariety of* T *generated by the* $\infty^{\frac{1}{2}(n^2-n)}$ *spaces* A; *and similarly the image points of all the* $\beta(q)$ *lie on a primal* $\mathscr{B}$ *of* T, *generated by the spaces* B. *We call* $\mathscr{A}$ *and* $\mathscr{B}$ *the* nuclear primals *of* T.

Proof. We first make two observations:

(i) In Ch. 3, § 5 we said that two augmented generators $\Pi_1(q_1)$ and $\Pi_2(q_2)$ were *compatible* if they belonged to the same Clifford regulus R, i.e. if Π_1 and Π_2 were the terminal generators and q_1, q_2 the terminal quadrics of R. Also we noted (Ch. 3, Prop. 5.3) that if γ_1, γ_2 are any two skew generators of Ω, and if $\gamma_1(q_1)$ is an augmentation of γ_1, then there exists a unique augmentation $\gamma_2(q_2)$ of γ_2 that is compatible with $\gamma_1(q_1)$.

(ii) As regards the parametrization of T by elements of the three matrices (1.2), we observe that this can be simplified by writing $A = C+S$, where C is symmetric and S is skew-symmetric. We then have $A^T = C-S$; whence, after removal of a factor of proportionality, the triad of matrices takes the form

$$[C-SC^{-1}S,\ 2SC^{-1},\ C^{-1}]. \tag{2.1}$$

The original n^2 indeterminates in A are now replaced by $\frac{1}{2}(n^2+n)$ new indeterminates in C and $\frac{1}{2}(n^2-n)$ in S. We note that (C, S) and $(-C, S)$ give the same point of T.

For the proof of Prop. 2.1 we may obviously confine our attention to one of the two types of augmented generator, say $\alpha(q)$.

Suppose then that Ω has equation $x^Ty = 0$ and that α is the generator of Ω given by $y = Sx$, where S is skew-symmetric. This implies, incidentally, that α is skew to the reference space Y.

If D is any non-singular symmetric matrix, then the cone $y^TDy = 0$ meets Y in a non-singular quadric q^* and thereby defines an augmented generator $Y(q^*)$ attached to Y. By (i) above, therefore, there exists a unique compatible augmented generator $\alpha(q)$ attached to α, and joined therefore to $Y(q^*)$ by a Clifford regulus R. We assert then that the equation of R is

$$y = (S+tD^{-1})\,x, \tag{2.2}$$

where t is the variable parameter. For plainly the regulus given by (2.2) contains Y and α; it is autopolar (and therefore Clifford), polar pairs in R being given by equal and opposite values of the parameter t; and it is easy to verify that its terminal quadric in Y is q^*. Moreover, its terminal quadric q in α is given by $x^TD^{-1}x = 0$. We note also, conversely, that every augmented generator attached to α is terminal in a unique Clifford regulus of the form (2.2).

To find the image point of $\alpha(q)$ on T, we replace C in (2.1) by tD^{-1} and then make t tend to zero; and, after removal of a factor of proportionality, we find that *the required image point (of the augmented generator $\alpha(q)$ given by $y = Sx$ with augmentation*

$x^T D^{-1} x = 0$) *is that whose coordinates are the elements of the matrices of the triad*

$$[-SDS,\ 2SD,\ D]. \tag{2.3}$$

This image point depends therefore, in all, on the values of n^2-1 parameters, namely on the $\frac{1}{2}(n^2-n)$ elements s_{ij} $(i<j)$ in S and on the $\frac{1}{2}(n^2+n)-1$ *ratios* of the elements $d_{ij}=d_{ji}$ of D.

If we keep the s_{ij} fixed, then the elements of the first and second of the three matrices (2.3) are only linear combinations of those of the third; and this means that all the augmented generators $\alpha(q)$ attached to a fixed α have image points which lie in a space A of dimension $\frac{1}{2}(n-1)(n+2)$ on T. If the s_{ij} are now also allowed to vary independently, then we see that the image points of all the $\alpha(q)$, for variable α, lie on a primal $\mathscr{A}$ of T, locus of the $\infty^{\frac{1}{2}(n^2-n)}$ spaces A, and similarly the image points of all the $\beta(q)$, for variable β, lie on another primal $\mathscr{B}$ of T. This completes the proof.

If $L(\mathscr{F})$ is the space on T which represents a linear system $\mathscr{F}$, then the augmented generators of $\mathscr{F}$ will be represented on $L(\mathscr{F})$ by points of its intersection with $\mathscr{A}$ or $\mathscr{B}$. The following Proposition describes the nature of this intersection.

PROPOSITION 2.2. *If $L(\mathscr{F})$ is the r-dimensional space on T which maps the polar pairs of an r-dimensional linear system $\mathscr{F}$ of Clifford parallels in S_{2n-1}, then the points of $L(\mathscr{F})$ which map the augmented generators belonging to $\mathscr{F}$ lie on a quadric primal $\mathscr{N}(\mathscr{F})$ of $L(\mathscr{F})$ – the* nuclear quadric *of the mapping of $\mathscr{F}$ on $L(\mathscr{F})$.*

Proof. Let $\mathscr{F}$ be given by the canonical equation

$$y = (\alpha I + \lambda_1 S_1 + \ldots + \lambda_{r-1} S_{r-1})\,x,$$

where the S_i satisfy the equations (3.1) of Ch. 4 for some value of k $(1 \leqslant k \leqslant r)$. For the relevant properties of such canonical equations we refer the reader to Ch. 4, § 4, and we add here the following remark:

(A) If $\lambda_1, \ldots, \lambda_{r-1}$ are regarded as independent indeterminates, then each of them can be expressed (though not necessarily uniquely) as a linear combination with constant coefficients of the elements of the matrix $\lambda_1 S_1 + \ldots + \lambda_{r-1} S_{r-1}$. (For otherwise the linear independence of $S_1, \ldots, S_{r-1}$ would be contradicted.)

To find the image point on T of the general polar pair (Π, Π') of $\mathscr{F}$ we substitute $A = \pm \alpha I + \lambda_1 S_1 + \ldots + \lambda_{r-1} S_{r-1}$ in (1.2) and make use of the relations (3.1) of Ch. 4 to simplify the result. We find then that the coordinates of the required image point are the elements of the matrices of the triad

$$[(\alpha^2 + \lambda_1^2 + \ldots + \lambda_{k-1}^2) I, 2\lambda_1 S_1 + \ldots + 2\lambda_{r-1} S_{r-1}, I].$$

By (A) above we can then make an allowable transformation to new coordinates $\xi_0, \ldots, \xi_N$ in the space of T so that the image point of (Π, Π') is given by

$$\left.\begin{aligned} \xi_0 = 1, \quad \xi_1 = \lambda_1, \ldots, \quad \xi_{r-1} = \lambda_{r-1}, \\ \xi_r = \alpha^2 + \lambda_1^2 + \ldots + \lambda_{k-1}^2, \quad \xi_{r+1} = \ldots = \xi_N = 0. \end{aligned}\right\} \tag{2.4}$$

For variable $\alpha, \lambda_1, \ldots, \lambda_{r-1}$, these are, as expected, the parametric equations of a linear space $L(\mathscr{F})$ on T. But now, putting $\alpha = 0$, we see that the subvariety in $L(\mathscr{F})$ which maps the augmented generators of $\mathscr{F}$ is the quadric $\mathscr{N}(\mathscr{F})$ given by

$$\xi_0 \xi_r = \xi_1^2 + \ldots + \xi_{k-1}^2, \quad \xi_{r+1} = \ldots = \xi_N = 0.$$

This completes the proof; but it should be noted that our analysis does not deal explicitly with those spaces of $\mathscr{F}$ which meet the reference space Y and which can only be represented in $L(\mathscr{F})$ by points which lie in the 'prime at infinity' of this space, given by $\xi_0 = 0$. We shall return to this later in § 5.

COROLLARY 2.2.1. *The nuclear quadric $\mathscr{N}(\mathscr{F})$ is non-singular if and only if $\mathscr{F}$ is non-singular ($k = r$). Otherwise, if $k < r$, $\mathscr{N}(\mathscr{F})$ is a quadric cone with an $[r-k-1]$ as vertex; and in particular, if $\mathscr{F}$ is totally singular ($k = 1$), $\mathscr{N}(\mathscr{F})$ breaks up into the prime at infinity in $L(\mathscr{F})$, given by $\xi_0 = 0$, together with another prime.*

This Corollary, it will be noted, provides a natural geometrical interpretation of the weight k of $\mathscr{F}$ and shows, in particular, that k is an invariant of $\mathscr{F}$ (cf. Ch. 4, § 4).

A Clifford regulus R within $\mathscr{F}$, as already noted (Th. 1.1), is mapped on $L(\mathscr{F})$ by a line, and we see now, from Prop. 2.2, that the terminal (augmented) generators of R correspond to the two points (necessarily distinct) in which this line meets the nuclear quadric $\mathcal{N}(\mathcal{F})$; in § 5 we shall see, conversely, that *every* proper chord of $\mathcal{N}(\mathcal{F})$ represents a Clifford regulus within $\mathcal{F}$. (A line which touches $\mathcal{N}(\mathcal{F})$ represents what may be called a *split* Clifford regulus, as described in Ex. 2 below.)

If $\mathcal{F}$ is singular, so that $\mathcal{N}(\mathcal{F})$ is a cone, then any space of $\mathcal{F}$ that is mapped on a point of the vertex of $\mathcal{N}(\mathcal{F})$ can clearly belong to no Clifford regulus within $\mathcal{F}$ and is therefore a singular space of $\mathcal{F}$ (cf. Ch. 4, Cor. 4.1.1).

As a final remark, we mention that there can exist what we may call *degenerate linear systems* $\mathcal{F}^*$ in S_{2n-1}, such that the spaces of a system $\mathcal{F}^*$ are mapped on T by the points of a flat space $L(\mathcal{F}^*)$ in which the nuclear quadric is a repeated prime. But since such systems $\mathcal{F}^*$ contain no Clifford reguli, they are of little geometric interest.

Ex. 1. Discuss the representation on the T-model of the system $\mathcal{G}$ of ∞^4 spaces given by equation (3.7) of Ch. 5, § 3, where S is a fixed non-singular skew-symmetric matrix of order $\frac{1}{2}n$ and $\alpha_0, \ldots, \delta_0$ are independent parameters. (It should be remembered that in the above context the equation of Ω was $x^T x' + y^T y' = 0$.) The coordinate ratios of the image point on T of a space of $\mathcal{G}$ all turn out to be linear combinations of

$$z_0 = 1,\ z_1 = \alpha_0,\ z_2 = \delta_0,\ z_3 = \beta_0 + \gamma_0 \text{ and } z_4 = \alpha_0\delta_0 - \beta_0\gamma_0,$$

so that $\mathcal{G}$ is a *linear* system of dimension 4. The augmented generators that belong to $\mathcal{G}$ occur when $\beta_0 = \gamma_0$, and they are therefore represented on T by points of the nuclear quadric $\mathcal{N}(\mathcal{G})$ with equation

$$z_3^2 - 4z_1 z_2 + 4z_0 z_4 = 0.$$

As this quadric is non-singular, it follows that $\mathcal{G}$ is also non-singular.

Show further that if the eigenvalues of

$$\begin{pmatrix} \alpha_0 & \beta_0 \\ \gamma_0 & \delta_0 \end{pmatrix}$$

are required to be mates in a fixed involution, then the corresponding spaces of $\mathcal{G}$ are represented on T by the points of a solid which does not touch $\mathcal{N}(\mathcal{G})$. These spaces therefore form a non-singular ∞^3 linear subsystem of $\mathscr{G}$, i.e. a system of weight 3. Interpret this result in the context of Ch. 5, § 3, with particular reference to Ex. 4 there following; and hence complete the proof of Th. 3.1 of Ch. 5.

Show finally that the spaces of $\mathscr{G}$ corresponding to matrices

$$\begin{pmatrix} \alpha_0 & \beta_0 \\ \gamma_0 & \delta_0 \end{pmatrix}$$

with two fixed unequal eigenvalues form a non-singular linear system of dimension 2.

Ex. 2. *Split Clifford reguli.* Let $\mathscr{F}$ be a linear ∞^r-system of Clifford parallels ($r > 1$) given by the canonical equation (3.2) of Ch. 4, let $\lambda_1, \ldots, \lambda_{r-1}$ be independent indeterminates and, as usual, let $\Lambda^2 = -(\lambda_1^2 + \ldots + \lambda_{k-1}^2)$, where k is the weight of $\mathscr{F}$. Suppose also that $k > 1$, so that $\Lambda \neq 0$ and n is necessarily even. Then the spaces $\Pi(t)$, $\Pi'(t)$ of $\mathscr{F}$ given respectively by

$$y = t(\Lambda I + \lambda_1 S_1 + \ldots + \lambda_{r-1} S_{r-1})\, x,$$

$$y = t(-\Lambda I + \lambda_1 S_1 + \ldots + \lambda_{r-1} S_{r-1})\, x,$$

are a pair of polar spaces with respect to Ω represented on the T-model (in the notation of equation (2.4)) by the point

$$\xi_0 = 1,\ \xi_1 = t\lambda_1, \ldots,\ \xi_{r-1} = t\lambda_{r-1},\ \xi_r = 0,\ \xi_{r+1} = \ldots = \xi_N = 0$$

and this point, as t varies, describes a generic tangent line to the nuclear quadric $\mathscr{N}(\mathscr{F})$ at the point $(1, 0, \ldots, 0)$.

The set of pairs $\Pi(t)$, $\Pi'(t)$, for given $\lambda_1, \ldots, \lambda_{r-1}$ and variable t, must therefore be regarded as a limiting kind of Clifford regulus – which we referred to above as a *split Clifford regulus* – whose terminal spaces have coincided. As an algebraic variety, such a split regulus has broken up into two halves ρ and ρ', each of dimension n and order $\frac{1}{2}n$, ρ being generated (for variable t) by $\Pi(t)$ and ρ' by $\Pi'(t)$. We shall not refer further to split reguli in this book and content ourselves with the following brief notes.

(i) The generators of ρ are the polars of those of ρ'.

(ii) The two halves have a generator of Ω as a common generator, and ρ and ρ' have no other common points. The common generator, like the terminal generators of a proper Clifford regulus, has a natural structure of an augmented generator.

(iii) Each half is a cone generated by the $[n-1]$'s joining a vertex $[\frac{1}{2}n-1]$ to an ordinary regulus of $[\frac{1}{2}n-1]$'s and lies therefore in a $[\frac{3}{2}n-1]$. The common generator mentioned in (ii) is the join of the vertices of these cones or, equally, the intersection of their ambient spaces.

For $n = 2$, let g be a generator of the absolute quadric surface Ω, let A, B be two points of g and let α, β be the tangent planes to Ω at A, B respectively. Then the pencils of lines $A(\beta)$ and $B(\alpha)$ are the two halves of a split Clifford regulus in S_3, the common (augmented) generator being $g(A, B)$.

Ex. 3. Show that the Clifford reguli that are contained in a linear system $\mathscr{F}$, given by its canonical equation in the usual form, are represented in the affine space of coordinates $\alpha, \lambda_1, \ldots, \lambda_{r-1}$ by conics of a certain family of which a general member has equations of the form

$$\alpha^2+\lambda_1^2+\ldots+\lambda_{k-1}^2 = a_1\lambda_1+b_1 = \ldots = a_{r-1}\lambda_{r-1}+b_{r-1},$$

where the a_i and b_j are arbitrary constants. Discuss the character of this family for low values of r.

3. Geometrical construction of T

Apart from its connection with the theory of Clifford parallels, the variety T is of some interest on its own account as a model of polar pairs (Π, Π') with respect to the quadric Ω. It would seem worth while, therefore, to investigate its geometrical character and construction more closely, and this and the following section are devoted to a digression on this topic. We shall show in this section that T is uniquely defined by a certain subvariety $\mathscr{V}$ of itself – its core variety – and that $\mathscr{V}$ is the quadratic Veronesean of a non-singular quadric Ω'_{2n-2}.

We begin by recalling from the previous section that if α is the generator $y = Sx$ of Ω, and if q is the quadric in which α is met by the cone $x^TD^{-1}x = 0$, where the matrix D is symmetric and non-singular, then the coordinates of the image point on T of the augmented generator $\alpha(q)$ are the elements of the matrices of the triad

$$[-SDS,\ 2SD,\ D]. \tag{3.1}$$

If the elements $s_{ij} = -s_{ji}$ of S and the elements $d_{ij} = d_{ji}$ of D are all taken to be independent indeterminates, then (3.1) can be regarded as the generic point of the nuclear primal $\mathscr{A}$ which is the image variety on T of all $\alpha(q)$; and $\mathscr{A}$ is generated by spaces A on T which each correspond to a replacement of the s_{ij} in (3.1) by a set of constants which are not all zero. Similarly, there exists on T a nuclear primal $\mathscr{B}$ which is the image variety of all $\beta(q)$; and the two primals $\mathscr{A}$ and $\mathscr{B}$ are generated respectively by the systems of spaces A and B on T which correspond to the individual generators α and β of Ω.

We now specialize D in (3.1) by taking it to be of rank 1, writing $D = \xi\xi^T$, where ξ is a column vector of indeterminates. Then (3.1) becomes

$$[-S\xi\xi^T S,\ 2S\xi\xi^T,\ \xi\xi^T]; \tag{3.2}$$

and if we write $\eta = S\xi$, this takes the form

$$[\eta\eta^T,\ 2\eta\xi^T,\ \xi\xi^T], \tag{3.3}$$

where ξ and η are subject only to the fixed relation

$$\xi^T\eta = 0. \tag{3.4}$$

If then we let Ω' be the quadric primal $x'^T y' = 0$ of a space S'_{2n-1} in which (x', y') are the current coordinates, we see that (ξ, η) is a generic point of Ω' and hence that (3.3) – subject to (3.4) – is a generic point of a subvariety $\mathscr{V}$ of T which is the quadratic Veronesean transform of Ω'. (To avoid confusion, we do not identify S'_{2n-1} and Ω' with S_{2n-1} and Ω; in fact, the matrix triad (3.3) is the coefficient image of the quadric

$$(x^T\eta + y^T\xi)(\eta^T x + \xi^T y) = 0$$

– an extreme specialization of the quadric $\mathscr{Q}(\Pi, \Pi')$ given by (1.1) – which consists of the repeated tangent prime to Ω at (ξ, η); thus the points of Ω' (or $\mathscr{V}$) are in natural correspondence with the tangent primes of Ω, so that Ω' is essentially the dual of Ω.) We shall refer to the subvariety $\mathscr{V}$ as the *core* of T, postponing till the following section its interpretation in relation to the T-representation.

The intersection of $\mathscr{V}$ with a generator A of $\mathscr{A}$, obtained by specializing S in (3.2) to a constant matrix, is a variety ϕ, of dimension $n-1$, which is the quadric Veronesean of an $[n-1]$-generator of one system of Ω' – say an α'-generator; and plainly A, being of dimension $\frac{1}{2}(n-1)(n+2)$, is the ambient space of ϕ. Thus $\mathscr{A}$ is now identified with the locus of ambient spaces A of the quadric Veroneseans ϕ on $\mathscr{V}$ which correspond to the α'-generators of Ω'. It follows now, by symmetry, that $\mathscr{B}$ is generated by the ambient spaces B of the quadric Veroneseans ψ on $\mathscr{V}$ which correspond to the β'-generators of Ω'. Hence:

(i) *The model T possesses a* core variety $\mathscr{V}$, *of dimension* $2n-2$, *which is the quadratic Veronesean of a non-singular quadric Ω' of S'_{2n-1}. The primals $\mathscr{A}$ and $\mathscr{B}$ of T (arising from the systems of $\alpha(q)$ and $\beta(q)$ of S_{2n-1}) are the loci of the ambient spaces A and B of the two systems of quadric Veroneseans ϕ and ψ on $\mathscr{V}$ which correspond to generators α' and β' of Ω'.*

It now remains to be shown that the core variety $\mathscr{V}$ uniquely determines not only $\mathscr{A}$ and $\mathscr{B}$ but also T itself. We propose to do this by showing that $\mathscr{V}$ of itself determines the whole family of lines of T (generating T multiply) which represent Clifford reguli of S_{2n-1}.

The argument is the same whether n is even or odd; in other words, whether the terminal generators of a Clifford regulus are of the same or of opposite systems on Ω. *We shall suppose, therefore, for convenience, that n is even.*

With this proviso, we consider a pair of skew generators α_1, α_2 of Ω and the corresponding generating spaces A_1 and A_2 of $\mathscr{A}$. A pair of points Q_1 and Q_2 of A_1 and A_2 respectively represent in general a pair of augmented generators $\alpha_1(q_1)$ and $\alpha_2(q_2)$; and the line Q_1Q_2 lies on T and represents a Clifford regulus if and only if $\alpha_1(q_1)$ and $\alpha_2(q_2)$ are compatible (cf. Ch. 3, § 5). Further, if α_1 and α_2 are fixed, the condition of compatibility implies that each of q_1, q_2 uniquely determines the other. There exists, therefore, a Cremona transformation between points Q_1, Q_2 of A_1 and A_2 respectively which represent compatible augmented generators $\alpha_1(q_1)$ and $\alpha_2(q_2)$.

To see the nature of this correspondence, we take α_1 and α_2 to be the generators X and Y of Ω. In this case, if q_2 in Y has equation $y^TDy = 0$, then its compatible q_1 in X has equation $x^TD'x = 0$

where $D' = D^{-1}$. The spaces A_1 and A_2 are those defined by the matrix triads $[0, 0, D]$ and $[D', 0, 0]$; and the Veroneseans ϕ_1 and ϕ_2 in which A_1 and A_2 meet $\mathscr{V}$ correspond to the generators $y' = 0$ and $x' = 0$ of Ω', their parametric representations being $(0, 0, \xi\xi^T)$ and $(\eta\eta^T, 0, 0)$ respectively. Then we may say:

(ii) *There exists, between points* $Q_1 = (d_{ij})$ *of* A_1 *and* $Q_2 = (d'_{ij})$ *of* A_2, *a symmetric determinantal Cremona transformation,*† *with matrix equation* $D' = D^{-1}$, *such that the lines* Q_1Q_2 *which join corresponding points lie on* T *and represent Clifford reguli of* S_{2n-1}.

Further, if $n > 2$, *the key base varieties of the above transformation are the Veroneseans* ϕ_1 *and* ϕ_2 *in which* A_1 *and* A_2 *meet* $\mathscr{V}$.

We shall suppose from now on that $n > 2$. The case $n = 2$, when the Cremona transformation is a plane collineation, is discussed separately below.

A Cremona transformation of the type described above, between spaces A_1 and A_2 which are each of dimension $\frac{1}{2}(n^2+n)-1$, will be called a $\tau(n)$; that is to say, the transformation must be such that, for suitable choice of the symmetrical arrays of coordinates d_{ij} in A_1 and d'_{ij} in A_2, it can be represented by the matrix equation $D' = D^{-1}$.

Suppose now that only the key base varieties ϕ_1 and ϕ_2 of such a $\tau(n)$ are prescribed, and that these have parametric representations, relative to fixed coordinate systems in A_1 and A_2, given by $(d_{ij}) = \xi\xi^T$ and $(d'_{ij}) = \eta\eta^T$. Then, if τ_0 is the transformation given by $D' = D^{-1}$, the equation of the general $\tau(n)$ with ϕ_1 and ϕ_2 as its key base varieties is of the form

$$MD'M^T = D^{-1}, \tag{3.5}$$

where M is any non-singular $n \times n$ matrix of constants; for this equation can be written in the form $D'' = D^{-1}$ by use of the coordinate transformation $D'' = MD'M^T$ in A_2, and it represents the product of τ_0 by a self-collineation of A_2 which leaves ϕ_2 invariant.

† This is, of course, the well-known Cremona transformation between the space of quadric loci of $[n-1]$ and the space of the corresponding quadric envelopes. The key base varieties are those representing the repeated primes of $[n-1]$ and the repeated points of $[n-1]$ respectively.

Now, for any $\tau(n)$ of the above kind, the neighbourhood of a point ξ of ϕ_1 is transformed into the ambient space of a sub-Veronesean variety ρ_2 of ϕ_2 – image of an $[n-2]$ of the $[n-1]$ of which ϕ_2 is the Veronesean; and similarly the neighbourhood of a point η of ϕ_2 is transformed into the ambient space of a sub-Veronesean ρ_1 of ϕ_1. It is easy to see then that both correspondences – from ξ to ρ_2 and from η to ρ_1 – arise from one and the same *correlation*† between ϕ_1 and ϕ_2; and that, for the $\tau(n)$ given by (3.5), the equation of this correlation is

$$\xi^T M \eta = 0.$$

Thus when ϕ_1 and ϕ_2 are assigned, a transformation $\tau(n)$ between A_1 and A_2, with ϕ_1 and ϕ_2 as its key base varieties, is uniquely determined by a non-singular correlation between ϕ_1 and ϕ_2 – that which determines the fundamental spaces in A_1 and A_2 which correspond to the neighbourhoods of points of ϕ_2 and ϕ_1. In particular, the correlation which determines τ_0 in this way is that given by $\xi^T \eta = 0$. Now, recalling that $\mathscr{V}$ is the quadratic Veronesean of Ω', and that ϕ_1 and ϕ_2 are the Veroneseans on $\mathscr{V}$ corresponding to the generators $y' = 0$ and $x' = 0$ of Ω', we notice that the correlation $\xi^T \eta = 0$ which determines τ_0 corresponds, on Ω', to the *natural correlation*‡ κ' which Ω' induces between $y' = 0$ and $x' = 0$.

More generally, if α_1' and α_2' are any two non-intersecting α'-generators on Ω', then Ω' sets up a natural correlation between α_1' and α_2'; to this there corresponds a uniquely defined correlation between their Veroneseans ϕ_1 and ϕ_2 on $\mathscr{V}$, and hence a unique transformation $\tau(n)$ between the ambient spaces A_1 and A_2 of ϕ_1 and ϕ_2, with ϕ_1 and ϕ_2 as the key base varieties; and the joins of points Q_1 and Q_2 in A_1 and A_2 which correspond by $\tau(n)$ lie on T and represent Clifford reguli in S_{2n-1}.

We can now sum up the results of this section in

THEOREM 3.1. *The T-model of pairs of polar $[n-1]$'s for Ω in S_{2n-1} is uniquely determined by its core variety $\mathscr{V}$, which is the*

† Since ϕ_1 and ϕ_2 are Veronesean transforms of two linear spaces, we may regard any correlation correspondence between the linear spaces as defining (by an abuse of language) a *correlation* between ϕ_1 and ϕ_2.

‡ Cf. Ch. 2, §2.

quadratic Veronesean of a non-singular quadric Ω' *of* S'_{2n-1}. *To the generators* α' *and* β' *of* Ω' *there correspond systems of Veroneseans* ϕ *and* ψ *on* $\mathscr{V}$, *and the ambient spaces* A *and* B *of these Veroneseans generate the nuclear primals* $\mathscr{A}$ *and* $\mathscr{B}$ *of* T *whose points correspond in general to the two classes* $\alpha(q)$ *and* $\beta(q)$ *of augmented generators of* Ω *in* S_{2n-1}. *Further,* $\mathscr{V}$ *determines a unique Cremona transformation* $\tau(n)$ *between each non-intersecting pair of the spaces* A *and* B – *of the same kind if* n *is even and of opposite kinds if* n *is odd* – *such that the joins of corresponding points in all such transformations* $\tau(n)$ *lie on* T *and represent Clifford reguli in* S_{2n-1}.

We now illustrate the above results by reference to the cases $n = 2$ and $n = 3$. Consideration of the exceptionally interesting case $n = 4$ will be postponed to Ch. 7, where it can be dealt with alongside other representations of the polar pairs (Π, Π') of S_7, such as arise from two–one projections of the Grassmannian $G(3, 7)$.

The case $n = 2$. The quadrics $\mathscr{Q}(\Pi, \Pi')$ in S_3 are readily identified as those which (*a*) meet Ω in a (skew) quadrilateral of generators and (*b*) are apolar to Ω. We find then that T is a $V_4^6[8]$, Segre product of a pair of planes π and π^*, and it lies in a prime of the image space S_9 of the quadrics of S_3. Its core variety $\mathscr{V}$, being the quadratic Veronesean of a quadric Ω' of S'_3, is a Del Pezzo surface ${}^1F^8$ of the second kind; and it arises on T as image of the product of a conic k in π by a conic k^* in π^*.

The varieties $\mathscr{A}$ and $\mathscr{B}$ on T are the threefolds generated by the two simply infinite families of conic planes A and B of ${}^1F^8$, and each is therefore a $V_3^6[8]$, Segre product of a conic by a plane. If S_1 and S_2 are any two conics of the first system on ${}^1F^8$, lying in planes A_1 and A_2, then there is a natural homographic correspondence between S_1 and S_2 determined by their intersections with conics of the second system on ${}^1F^8$, and this homography is subordinate to a unique collineation $\tau(2)$ between A_1 and A_2 such that the joins of corresponding points lie on V_4^6 and represent Clifford reguli in S_3.

The case $n = 3$. The variety T in this case, as already stated (§ 1, Ex. 4), is the variety V_9^{92}, otherwise studied as the model of quadric complexes of S_3. Its core variety $\mathscr{V}$ is the quadratic

Veronesean of a quadric Ω' of S'_5; the Veronese surfaces on $\mathscr{V}$ corresponding to the generating planes α' and β' of Ω' are such that their ambient [5]'s A and B generate the nuclear† primals $\mathscr{A}$ and $\mathscr{B}$ on T; and if ϕ_0 and ψ_0 are any two of these Veronese surfaces, lying in spaces A_0 and B_0 respectively, then the natural correlation between ϕ_0 and ψ_0 is determined by the Veronese surfaces on $\mathscr{V}$ which meet one of them in a point and the other in a conic, and this correlation is uniquely associated with a Cremona transformation $\tau(3)$ of A_0 into B_0 which carries primes of A_0 into quadrics through ψ_0 and quadrics through ϕ_0 into primes of B_0.

Ex. 1. *The Study representation for $n = 2$.* In connection with the T-model for $n = 2$, as we have described it above, it is of interest to note that an equivalent of this representation is already known in the form of Study's direct mapping‡ of the (unordered) pairs of polar lines (p, p') with respect to a quadric surface Ω in S_3 on the pairs of points (P, P^*) of two planes π and π^*. The construction for this is as follows. Let O be a fixed point not on Ω, and let π and π^* be two fixed planes not through O. Further let the generators of the first system on Ω that meet any general line p be α_1, α_2, and let those of the second system that meet p be β_1, β_2. If the transversals from O to the pairs (α_1, α_2) and (β_1, β_2) meet π and π^* in P and P^* respectively, then we say that to p there corresponds the point-pair (P, P^*); and we observe that the polar line p' of p (and only this line) corresponds in this way to the same point-pair (P, P^*). We have, in fact, a birational mapping of all the (unordered) polar pairs (p, p') on the (ordered) point-pairs (P, P^*) of π and π^*. The connection with our previous mapping will now be apparent, the conics k and k^* of that mapping being identifiable with the conics in which the tangent cone from O to Ω meets the planes π and π^*.

The above representation has the outstanding property that if two polar pairs (p, p') and (q, q') are such that each of p and p'

† Semple[21] has shown that V_9^{92} passes sextuply through the 4-fold $\mathscr{V}$ and has no other singularity. The two varieties $\mathscr{A}$ and $\mathscr{B}$ each pass sextuply through $\mathscr{V}$ and they intersect simply in a 7-fold $\mathscr{A}\mathscr{B}$ which passes octuply through $\mathscr{V}$.

‡ Cf. Coxeter[6], p. 146.

meets each of q and q' (i.e. if each of p and p' is perpendicular, in the non-Euclidean sense, to each of q and q'), then their representative point-pairs (P, P^*) and (Q, Q^*) are such that P and Q are conjugate for k, while P^* and Q^* are conjugate for k^*. By this property, any theorem concerning pairs of conjugate points with respect to a conic gives rise to a non-Euclidean theorem concerning perpendicular lines. For example, Hesse's theorem (that a triangle is in perspective with its polar triangle with respect to a given conic) gives rise to a non-Euclidean theorem of which the Euclidean specialization is the well-known Petersen–Morley theorem (that the three common perpendicular transversals of pairs of opposite sides of a skew rectangular hexagon possess a common perpendicular transversal).

4. Fundamental subvarieties of T

The core variety $\mathscr{V}$ of T, as stated in Th. 3.1, is the quadratic Veronesean of a non-singular quadric Ω' of S'_{2n-1}, and the ambient spaces A and B of the Veroneseans ϕ and ψ on $\mathscr{V}$ which correspond to generators α' and β' of Ω' generate the two fundamental primals $\mathscr{A}$ and $\mathscr{B}$ of T. There are, however, lesser fundamental varieties on T generated by the ambient spaces of those Veroneseans on $\mathscr{V}$ that correspond to lines, planes, ..., $[n-2]$'s on Ω'. We denote these by $\mathscr{V}^{(1)}, \mathscr{V}^{(2)}, \ldots, \mathscr{V}^{(n-2)}$, regarding $\mathscr{V}$ itself as $\mathscr{V}^{(0)}$; and we then have the inclusion diagram

$$\mathscr{V} = \mathscr{V}^{(0)} \subset \mathscr{V}^{(1)} \subset \ldots \subset \mathscr{V}^{(n-2)} \begin{matrix} \subset \mathscr{A} \\ \subset \mathscr{B} \end{matrix} \subset T.$$

In other words, $\mathscr{V}^{(1)}$ is the locus of conic-planes of $\mathscr{V}$, $\mathscr{V}^{(2)}$ is the locus of [5]'s which meet $\mathscr{V}$ in Veronese surfaces, and so on. We note especially that $\mathscr{V}^{(n-2)}$ is the intersection of $\mathscr{A}$ and $\mathscr{B}$. We now wish to explain the role of these $\mathscr{V}^{(i)}$ in the birational correspondence ρ between the polar pairs (Π, Π') of S_{2n-1} and the points of T.

Since T itself was defined as the variety (locus of specializations) of a certain generic point (cf. (1.2)), and since this point was the coefficient image of the quadric $\mathscr{Q}(\Pi, \Pi')$ with equation (1.1), it follows that T is the coefficient image of an irreducible system of quadrics – to be denoted by $(\mathscr{Q})$ – defined by its generic member $\mathscr{Q}(\Pi, \Pi')$; and we may equally regard ρ as the birational correspondence between polar pairs (Π, Π') and quadrics of the system $(\mathscr{Q})$. Then $(\mathscr{Q})$ will have a set of fundamental subsystems $(\mathscr{Q}^{(0)}), (\mathscr{Q}^{(1)}), \ldots, (\mathscr{Q}^{(n-2)}), (\mathscr{Q}^{(n-1)}_{\mathscr{A}})$ and $(\mathscr{Q}^{(n-1)}_{\mathscr{B}})$ corresponding to the fundamental varieties $\mathscr{V}^{(0)}, \mathscr{V}^{(1)}, \ldots, \mathscr{V}^{(n-2)}, \mathscr{A}$ and $\mathscr{B}$ of T. We now assert:

(i) $(\mathscr{Q}^{(n-1)}_{\mathscr{A}})$ *is the system of quadric cones of* S_{2n-1} *which have an* α*-generator of* Ω (*or a space containing an* α*-generator*) *as vertex; and* $(\mathscr{Q}^{(n-1)}_{\mathscr{B}})$ *is defined similarly;*

(ii) *if* $0 \leqslant h \leqslant n-2$ *and* $k = 2n-2-h$, *then* $(\mathscr{Q}^{(h)})$ *is the system of quadric cones of* S_{2n-1} *which have the polar space* S'_k *of any* S_h *of* Ω *as vertex* (*or whose vertices contain such a space* S'_k); *and, in particular,*

(iii) $(\mathscr{Q}^{(0)})$ *is the system of repeated tangent primes of* Ω.

The significance for the T-representation of the varieties $\mathscr{V}^{(h)}$ $(h = 0, \ldots, n-2)$ together with $\mathscr{A}$ and $\mathscr{B}$ can now be stated as follows:

(iv) *For* $h = 0, \ldots, n-2$, $\mathscr{V}^{(h)}$ *is the image variety on* T *of all those specializations of the generic* $\mathscr{Q}$ *that correspond under* ρ *to polar pairs* (Π, Π') *such that* Π *meets* Π' *in a space* S_h *of* Ω. *Similarly* $\mathscr{A}$ *and* $\mathscr{B}$ *are the image varieties on* T *of those specializations of* $\mathscr{Q}$ *that correspond under* ρ *to coincident polar pairs* (α, α) *and* (β, β) *respectively.*

It should be understood, of course, that if Π_0 and Π'_0 meet in an S_h of Ω, and if $\bar{P}$ is the image on T of a $\bar{\mathscr{Q}}$ which is one of those arising under ρ from (Π_0, Π'_0), then (iv) only tells us that $\bar{P}$ lies in a certain generator of $\mathscr{V}^{(h)}$. Its position in this generator will depend on the 'approach path' from the generic (Π, Π') to (Π_0, Π'_0). Also the same point $\bar{P}$ may correspond under ρ to other polar pairs (Π_1, Π'_1) such that Π_1 meets Π'_1 in S_h.

We shall now indicate in outline how the above statements (i), …, (iv) can be verified, and we begin by proving

PROPOSITION 4.1. *If* γ *is any* $[n-1]$*-generator of* Ω, *then every quadric cone with vertex* γ (*or with vertex containing* γ) *is a specialization of the generic* $\mathscr{Q}(\Pi, \Pi')$.

Proof. By equation (2.1) of § 2, the equation of the generic $\mathcal{Q}(\Pi, \Pi')$ can be written in the form

$$x^T(C - SC^{-1}S)\,x + 2x^T SC^{-1}y + y^T C^{-1} y = 0, \qquad (4.1)$$

where the matrices C and S are symmetric and skew-symmetric respectively, with independent indeterminates as elements. Let γ be the generator $y = S_0 x$, and let $C+S$ be specialized to $tD^{-1} + S_0$, where D is a symmetric matrix of indeterminates and t is a further indeterminate. From (4.1), the resulting specialization $\overline{\mathcal{Q}}$ of $\mathcal{Q}(\Pi, \Pi')$ has equation

$$t^2 x^T D^{-1} x + \{-x^T S_0 D S_0 x + 2x^T S_0 D y + y^T D y\} = 0;$$

and when $t \to 0$, we get the further specialization $\mathcal{Q}^*$ given by

$$-x^T S_0 D S_0 x + 2x^T S_0 D y + y^T D y = 0,$$

which can be written in the form

$$(y - S_0 x)^T D (y - S_0 x) = 0. \qquad (4.2)$$

Thus (4.2) is a specialization of $\mathcal{Q}(\Pi, \Pi')$, and since it represents the generic cone with vertex γ, the proof of the Proposition is complete.

Since we proved earlier that $\mathcal{A}$ (resp. $\mathcal{B}$) is the fundamental variety of T which corresponds by ρ to all coincidence polar pairs (α, α) (resp. (β, β)), Prop. 4.1 is equivalent to statement (i).

We now prove

PROPOSITION 4.2. *If* (Π, Π') *is a generic polar pair and* (Π_0, Π_0') *is a polar pair such that* Π_0 *meets* Π_0' *in a space* S_h $(0 \leqslant h \leqslant n-1)$ *on* Ω, *then as* $(\Pi, \Pi') \to (\Pi_0, \Pi_0')$ *in any manner, the quadric* $\mathcal{Q}(\Pi, \Pi') \to$ *a quadric cone* $\overline{\mathcal{Q}}$ *with the join of* Π_0 *to* Π_0' (*or a space containing this join*) *as vertex.*

Proof. Since S_h lies on Ω, the join of Π_0 to Π_0' is the polar space S_k' of S_h, where $k = 2n-2-h$.

Let the sections of Ω by Π and Π' be q and q', and let K_1 and K_2 be the cones $\Pi(q')$ and $\Pi'(q)$. Then $\mathcal{Q}(\Pi, \Pi') \equiv \mathcal{Q}$ is the harmonic conjugate of Ω with respect to K_1 and K_2 in the pencil they determine. Now let $(\Pi, \Pi') \to (\Pi_0, \Pi_0')$ in any manner, and let the corresponding specializations of K_1 and K_2 be $\overline{K}_1$ and $\overline{K}_2$.

Then since Π is a double locus on K_1, Π_0 will be double on $\bar{K}_1$; and similarly Π_0' will be a double locus on $\bar{K}_2$. Thus, in particular, S_h will be double on each of $\bar{K}_1$ and $\bar{K}_2$. Now, since K_1, K_2 and Ω are linearly dependent, so in the limit must $\bar{K}_1$, $\bar{K}_2$ and Ω be linearly dependent. If $\bar{K}_1$ and $\bar{K}_2$ were distinct, then the linear dependence would require that S_h, being double for both, would also be double for Ω – an impossibility since Ω is non-singular. Hence $\bar{K}_1$ and $\bar{K}_2$ coincide, and we write $\bar{K}_1 \equiv \bar{K}_2 = \bar{K}$.

Further, since $\mathcal{Q}$ and Ω separate K_1 and K_2 harmonically, it follows now that the limit $\bar{\mathcal{Q}}$ of $\mathcal{Q}$, since it must be the harmonic conjugate of Ω with respect to $\bar{K}_1$ and $\bar{K}_2$, is $\bar{K}$. Also, since $\bar{K}$, regarded as $\bar{K}_1$, will have Π_0 as double locus, and, regarded as $\bar{K}_2$, will have Π_0' as double locus, it follows that $\bar{K}$ (i.e. $\bar{\mathcal{Q}}$) has the join of Π_0 and Π_0' as double locus.

Hence, as $(\Pi, \Pi') \to (\Pi_0, \Pi_0')$, the resulting specialization $\bar{\mathcal{Q}}$ of $\mathcal{Q}(\Pi, \Pi')$ is a quadric with the join S_k' of Π_0 and Π_0' (or a space containing this join) as vertex. This completes the proof.

To carry the argument further, we recall from § 3 (cf. equations (3.3) and (3.4) and the remarks there following) that the points of the core variety $\mathscr{V}$ of T are the coefficient images of the repeated tangent primes of Ω, the latter being extreme specializations of the generic $\mathcal{Q}(\Pi, \Pi')$. Thus to any S_h on Ω, i.e. to the aggregate of repeated tangent primes of Ω at points of S_h, there corresponds on $\mathscr{V}$ an h-dimensional quadric Veronesean† V_h whose ambient space is a $[\frac{1}{2}h(h+3)]$. Now to any generator α of Ω that passes through S_h there corresponds by ρ a space A; and this space A must clearly contain V_h and hence also its ambient $[\frac{1}{2}h(h+3)]$. It readily appears, then, that *this* $[\frac{1}{2}h(h+3)]$ *is the intersection of all the spaces A which correspond to α-generators of Ω through S_h.*

Now consider again the quadric $\bar{\mathcal{Q}} = \bar{K}$ which arose in the proof of Prop. 4.2 as a specialization of the generic $\mathcal{Q}$ arising from a polar pair (Π_0, Π_0') such that Π_0 meets Π_0' in S_h. Since this $\bar{\mathcal{Q}}$ had vertex S_k' – the polar of S_h – it follows that every α-generator through S_h will lie in S_k'; and hence $\bar{\mathcal{Q}}$ can be regarded (for each

† The quadric Ω' (of which $\mathscr{V}$ is the quadratic Veronesean) is, as we have seen in §3, in dual correspondence with Ω, the tangent primes of Ω being in natural correspondence with the points of Ω'. Hence the tangent primes of Ω at points of S_h correspond on Ω' to the points of an $[h]$, and V_h is the quadric Veronesean of this $[h]$.

such α) as a cone with that α-generator as vertex. Hence *the image point of* $\overline{\mathcal{Q}}$ *on* T *must lie in the intersection of all the spaces* A *that correspond to* α*-generators through* S_h.

Combining now the last two observations, and noting also that the freedom of quadric cones with S'_k as vertex is $\frac{1}{2}h(h+3)$, we deduce that all the $\overline{\mathcal{Q}}$ arising from polar pairs which intersect in S_h are represented on T by the points of the ambient $[\frac{1}{2}h(h+3)]$ of the quadric Veronesean V_h on $\mathscr{V}$. Thus, letting S_h vary on Ω, it follows that the image variety on T of all specializations of the generic $\mathcal{Q}$ that arise from polar pairs which intersect in an S_h is the fundamental variety $\mathscr{V}^{(h)}$ of T. This completes, therefore, our verification of the statements (i), ..., (iv) which we made at the beginning of this section.

5. Self-collineations of a linear system $\mathscr{F}$

In this section we take up again from § 2 the discussion of an r-dimensional linear system $\mathscr{F}$ of Clifford parallels of S_{2n-1} and the mapping there described of the polar pairs (Π, Π') of $\mathscr{F}$ on the points of a linear space $L(\mathscr{F})$ on the model T. We then use this mapping to discuss the group Γ of permutations on the spaces of $\mathscr{F}$ that are induced by self-collineations of S_{2n-1} which leave Ω and $\mathscr{F}$ invariant. From this discussion there will emerge certain structural homogeneity properties of those linear systems $\mathscr{F}$ which we have termed 'non-singular', emphasizing their relative simplicity and importance.

We recall, then, the previous mapping of the polar pairs (Π, Π') of $\mathscr{F}$ on the points of $L(\mathscr{F})$ as follows. We took the canonical equation of $\mathscr{F}$, subject to the conditions (3.1) of Ch. 4, in the form

$$y = (\alpha I + S(\lambda))\, x,$$

where

$$S(\lambda) = \sum_{i=1}^{r-1} \lambda_i S_i$$

and we saw then that a coordinate system $\xi_0, \ldots, \xi_r$ in $L(\mathscr{F})$ could be chosen in such a way that the polar pair given by $y = (\pm\alpha I + S(\lambda))\, x$ was represented in $L(\mathscr{F})$ by the point

$$(\xi_0, \ldots, \xi_r) = (1, \lambda_1, \lambda_2, \ldots, \lambda_{r-1}, \alpha^2 + \lambda_1^2 + \ldots + \lambda_{k-1}^2).$$

This implies, in particular, that the spaces of $\mathscr{F}$ which meet the reference space Y can only be represented in $L(\mathscr{F})$ by points lying in the 'prime at infinity' $\xi_0 = 0$ in this space. Further, we saw that the augmented generators of $\mathscr{F}$, based on generators of Ω with equations of the form $y = S(\lambda)x$, were represented in $L(\mathscr{F})$ by points of the quadric $\mathscr{N}(\mathscr{F})$ with equation

$$\xi_0 \xi_r = \xi_1^2 + \ldots + \xi_{k-1}^2.$$

We note here also that of all the augmented generators based on a generator $y = S(\lambda)x$ contained in $\mathscr{F}$ only one – the *augmented generator properly belonging to* $\mathscr{F}$ (cf. Ch. 4, remark following Cor. 3.1.1) – is represented in $L(\mathscr{F})$ by a point, and this point lies on $\mathscr{N}(\mathscr{F})$.

Now consider the permutations on the spaces of $\mathscr{F}$ that arise from self-collineations of S_{2n-1} which leave Ω and $\mathscr{F}$ invariant. We begin with the remark:

Any self-collineation of S_{2n-1} which leaves Ω invariant – a (complex) non-Euclidean 'congruence transformation' – induces a self-collineation of T.

For, in the first place, any self-collineation ϖ of S_{2n-1} induces a self-collineation ϖ' of the space S_N of quadric primals of S_{2n-1}; and if ϖ leaves Ω invariant, so that it carries polar pairs always into polar pairs, then ϖ' must carry T (a variety in S_N) into itself.

From this it follows that if ϖ also leaves $\mathscr{F}$ invariant, then ϖ' must leave $L(\mathscr{F})$ invariant, i.e. it induces a self-collineation of $L(\mathscr{F})$; and further, since ϖ permutes the (augmented) generators in $\mathscr{F}$ among themselves, it follows that ϖ' must transform the quadric $\mathscr{N}(\mathscr{F})$ into itself. This proves

PROPOSITION 5.1. *If $\mathscr{F}$ is any linear system of parallels in S_{2n-1}, then every self-collineation of S_{2n-1} which leaves Ω and $\mathscr{F}$ invariant induces a self-collineation of $L(\mathscr{F})$ which leaves the quadric $\mathscr{N}(\mathscr{F})$ invariant.*

It is important to notice that the self-collineation of $L(\mathscr{F})$ may be the identity, even when the self-collineation of S_{2n-1} from which it derives is not. Thus there is a difference between the group of self-collineations of S_{2n-1} which leave Ω and $\mathscr{F}$ invariant and the group of permutations of the spaces of $\mathscr{F}$ induced by such self-collineations.

Now consider any chordal polar pair (Π_0, Π_0') in S_{2n-1}, and let $\varpi(\Pi_0, \Pi_0')$ denote the biaxial harmonic homology in S_{2n-1} with Π_0 and Π_0' as axial spaces – a type of collineation which we may refer to as a *skew reflection* with respect to Ω. Plainly $\varpi(\Pi_0, \Pi_0')$ leaves Ω invariant and carries polar pairs into polar pairs. Suppose then that (Π_0, Π_0') belongs to $\mathscr{F}$. Then if (Π, Π') is a generic polar pair of $\mathscr{F}$, there exists a Clifford regulus R containing (Π_0, Π_0') and (Π, Π') and wholly contained in $\mathscr{F}$. Hence $\varpi(\Pi_0, \Pi_0')$ transforms R into itself, permuting the polar pairs of R among themselves (cf. Ch. 3, § 4, Ex. 1). Hence:

If (Π_0, Π_0') *is any polar chordal pair of* $\mathscr{F}$, *then the skew reflection with* Π_0, Π_0' *as axial spaces leaves* Ω *and* $\mathscr{F}$ *invariant, permuting polar pairs of* $\mathscr{F}$ *among themselves.*

We now prove

PROPOSITION 5.2. *If* (Π_0, Π_0') *is any chordal polar pair of a linear system* $\mathscr{F}$, *then the skew reflection* $\varpi(\Pi_0, \Pi_0')$ *induces in* $L(\mathscr{F})$ *a central harmonic homology* $\varpi'(P_0)$ *whose centre* P_0 *is the image point of* (Π_0, Π_0') *and whose axial prime is the polar of* P_0 *with respect to the quadric* $\mathscr{N}(\mathscr{F})$.

Plainly $\varpi'(P_0)$ leaves $\mathscr{N}(\mathscr{F})$ invariant, and we shall call it a *central reflection* in $L(\mathscr{F})$ with respect to $\mathscr{N}(\mathscr{F})$.

Proof. Let $\varpi(\Pi_0, \Pi_0')$ carry a generic polar pair (Π_1, Π_1') of $\mathscr{F}$ into (Π_2, Π_2'). Then, by what we have said above, all the six spaces concerned belong to a Clifford regulus R contained in $\mathscr{F}$; and we can choose a coordinate system (Ch. 3, Th. 4.2) in which Ω has equation $x^T y = 0$ and R is given by $y = \theta x$. Let Π_0, Π_0' be given in R by $\theta = \pm\alpha$. Then (cf. Ch. 3, § 4, Ex. 1) $\varpi(\Pi_0, \Pi_0')$ carries $y = \theta x$ into $y = (\alpha^2/\theta)x$; so that, if Π_1, Π_1' are given by $y = \pm\theta_1 x$, then Π_2, Π_2' are given by $y = \pm(\alpha^2/\theta_1)x$. Now, on the T-model, the pairs of polar spaces $y = \pm\theta x$ are represented by the points of a line on which $\phi = \theta^2$ is a projective parameter. The pair (Π_0, Π_0') is represented on this line by the point P_0 for which $\phi = \alpha^2$, the pair (Π_1, Π_1') by the point P_1 given by $\phi = \theta_1^2$; and the pair (Π_2, Π_2') by the point P_2 for which $\phi = \alpha^4/\theta_1^2$. Now the correspondence between P_1 and P_2, given by $\phi \to \alpha^4/\phi$, is involutory with united points given by $\phi = \pm\alpha^2$. One of these united points $(\phi = \alpha^2)$ is P_0, and the other $(\phi = -\alpha^2)$ is the harmonic conjugate of P_0 with respect to the points given by $\phi = 0$ and $\phi = \infty$, these latter being evidently the two points of $\mathscr{N}(\mathscr{F})$ which represent the terminal spaces of R (given by $\theta = 0$ and $\theta = \infty$). In other words, the harmonic conjugate of P_0 with respect to P_1 and P_2 lies on the polar of P_0 with respect to $\mathscr{N}(\mathscr{F})$. Thus the transformation carrying P_1 into P_2 is the central reflection (relative to $\mathscr{N}(\mathscr{F})$) with P_0 as vertex. This completes the proof.

For the special case in which the linear system $\mathscr{F}$ is non-singular, the result just proved has certain important implications. In this case, as already remarked, $\mathscr{N}(\mathscr{F})$ is a non-singular quadric in $L(\mathscr{F})$; and hence, by a known theorem,† the central reflections in $L(\mathscr{F})$ with respect to $\mathscr{N}(\mathscr{F})$ generate the whole group Γ' of self-collineations of $L(\mathscr{F})$ which leave $\mathscr{N}(\mathscr{F})$ invariant. Hence, when $\mathscr{F}$ is non-singular, every self-collineation of $\mathscr{N}(\mathscr{F})$ corresponds to a permutation of the spaces of $\mathscr{F}$ which is generated by a finite sequence of the skew reflections $\varpi(\Pi, \Pi')$ of S_{2n-1} such that (Π, Π') is a chordal polar pair of $\mathscr{F}$. Also, conversely, by Prop. 5.1, every permutation of the spaces of $\mathscr{F}$ which is generated by a self-collineation of S_{2n-1} leaving Ω and $\mathscr{F}$ invariant corresponds to a self-collineation of $L(\mathscr{F})$ belonging to the group Γ'. Hence:

COROLLARY 5.2.1. *When $\mathscr{F}$ is non-singular, the group Γ of permutations of the spaces of $\mathscr{F}$ arising from self-collineations of S_{2n-1} which leave Ω and $\mathscr{F}$ invariant is isomorphic with the group Γ' of self-collineations of $L(\mathscr{F})$ which leave $\mathscr{N}(\mathscr{F})$ invariant.*

But now further (still supposing $\mathscr{F}$ to be non-singular), we remark that the group Γ' is transitive at two levels on the points of $L(\mathscr{F})$, distinguishing only between points of $L(\mathscr{F})$ which do or do not lie on $\mathscr{N}(\mathscr{F})$. Thus in particular, if P is any point in the 'prime at infinity' $\xi_0 = 0$ of our previous construction, there exists a self-collineation of Γ' which carries P into a 'finite' point

† Cf. Burau [4], p. 234.

P', image either of a chordal polar pair or of an augmented generator of $\mathscr{F}$. It follows therefore, since either of these latter is carried by any permutation of Γ into another of the same kind, that every point such as P (like every 'finite' point of $L(\mathscr{F})$) corresponds either to a chordal polar pair or to an augmented generator of $\mathscr{F}$. This gives

Corollary 5.2.2. *If $\mathscr{F}$ is non-singular, then all the points of $L(\mathscr{F})$, including those which lie in the 'prime at infinity' $\xi_0 = 0$, represent either chordal pairs or augmented generators of $\mathscr{F}$, and $\mathscr{F}$ consists of such spaces and no others. The permutation group Γ of $\mathscr{F}$ is transitive on each of the two types of spaces of $\mathscr{F}$.*

The group Γ', we now note, is also transitive on the lines of $L(\mathscr{F})$ which meet $\mathscr{N}(\mathscr{F})$ in two distinct points. It follows that *every* such line represents a Clifford regulus within $\mathscr{F}$ and, in particular, that every space of $\mathscr{F}$ belongs to some Clifford regulus within $\mathscr{F}$. Hence, by Ch. 4, Cor. 4.1.1, we deduce that a non-singular system $\mathscr{F}$ cannot contain any singular space. This (in conjunction with the remark concerning singular spaces after Cor. 2.2.1 of this Chapter) proves

Corollary 5.2.3. *A linear system $\mathscr{F}$ is non-singular if and only if it contains no singular spaces.*

Finally, for a non-singular system $\mathscr{F}$, we note that if Σ is a terminal (augmented) generator of any Clifford regulus within $\mathscr{F}$, represented in $L(\mathscr{F})$ by a point P of $\mathscr{N}(\mathscr{F})$, then the Σ-affine part of $\mathscr{F}$ (as defined in Ch. 4, § 4) is represented by the points of $L(\mathscr{F})$ that do not lie in the tangent prime to $\mathscr{N}(\mathscr{F})$ at P. Now if $\mathscr{F}$ has freedom $r > 1$, and if P_1 and P_2 are two arbitrary points of $L(\mathscr{F})$, we can always find a point P of $\mathscr{N}(\mathscr{F})$ such that the tangent prime to $\mathscr{N}(\mathscr{F})$ at P does not pass through P_1 or P_2. This implies then that, given two arbitrary spaces Π_1, Π_2 of $\mathscr{F}$, it is always possible to find a Σ (terminal in some Clifford regulus of $\mathscr{F}$) such that Π_1 and Π_2 belong to the Σ-affine part of $\mathscr{F}$. Hence, by Prop. 4.2 of Ch. 4, we deduce

Corollary 5.2.4. *If two distinct spaces of a non-singular linear system intersect, then their intersection lies on Ω.*

Corollaries 5.2.3 and 5.2.4, it will be noted, give final confirmation to two results which were mentioned without proof in Ch. 4, § 4.

If $\mathscr{F}$ is a singular linear system, then $\mathscr{N}(\mathscr{F})$ is a cone; the central reflections with respect to $\mathscr{N}(\mathscr{F})$ no longer generate the whole group Γ' of self-collineations of $\mathscr{N}(\mathscr{F})$, and it need no longer be true that all the spaces of $\mathscr{F}$ are either chordal or augmented generators. In other words, the prime $\xi_0 = 0$ may contain points – possibly fundamental for the T-representation – which represent spaces of $\mathscr{F}$ that are neither chordal nor augmented generators; and in fact we have already noted the existence of singular systems which possess spaces of this kind (Ch. 4, § 4, Ex. 4). On the other hand if a singular system $\mathscr{F}$ is such that it is contained in a more ample non-singular linear system $\mathscr{F}'$, then the first conclusion of Cor. 5.2.2, being valid for $\mathscr{F}'$, is necessarily also valid for $\mathscr{F}$.

The relatively simple structure of non-singular linear systems, as set out in Cor. 5.2.2, is the main justification for the emphasis we have placed on them throughout this book.

Ex. 1. If (Π_0, Π_0') is the polar pair given by $y = (\pm \alpha I + S(\lambda))\, x$, where $S(\lambda) = \sum_{i=1}^{r-1} \lambda_i S_i$, and if Π is the $[n-1]$ given by

$$y = (\beta I + S(\mu))\, x,$$

show that the transform Π_1 of Π by the skew-reflection $\varpi(\Pi_0, \Pi_0')$ has equation

$$y = \phi^{-1}\{\beta I + (\phi + 1)\, S(\lambda) - S(\mu)\}\, x$$

where
$$\phi = \frac{1}{\alpha^2}\left\{\beta^2 + \sum_{i=1}^{k-1} (\mu_i - \lambda_i)^2\right\}.$$

CHAPTER 7

HALF-GRASSMANNIANS

In this final chapter we discuss a representational method which is different from that described in Ch. 6, employing two new models of the polar pairs (Π, Π') of S_{2n-1}; and these new models will be called half-Grassmannians because they are, in effect, (2, 1)-projections of the Grassmannian $G(n-1, 2n-1)$ of all $[n-1]$'s of S_{2n-1}. After a rapid survey of the general theory of this method and brief discussions of the cases $n = 2$ and $n = 3$, we treat in some detail the case $n = 4$ which is in many ways unique. This leads us directly to a simple construction and interpretation of the so-called Study triality correspondence between the points, α-solids and β-solids of a quadric in S_7. Only in this case ($n = 4$) does it happen that each of the two half-Grassmannians concerned is projectively equivalent to the corresponding T-model; and the birational (but non-linear) correspondences between pairs of the three models that result from the three different representations give rise to an extended form of the Study triality.

1. Grassmannian representation

We denote by G the Grassmannian $G(n-1, 2n-1)$ of all $[n-1]$'s of S_{2n-1}, noting that G is of dimension n^2 and that its ambient space is S_N, where $N = \binom{2n}{n} - 1$.

Further, we denote by G_α and G_β the subvarieties of G that map the systems of α-generators and β-generators of Ω. The properties of a variety such as G_α or G_β, which represents the generators of one system of a quadric in S_{2n-1}, are well known.† Thus, for example, G_α has a minimum model M_α of which it is precisely the quadratic Veronesean. In particular for $n = 2$, M_α is a line and G_α is a conic; for $n = 3$, M_α is a solid and G_α is a

† See, for example, Burau [**4**], Ch. 9 and Heymans [**11, 12**].

Veronesean $V_3^8[9]$; while for $n = 4$, M_α is a quadric Q_6^2 of S_7 and G_α is its quadratic Veronesean† $V_6^{128}[34]$. For the moment we note only that G_α and G_β are each of dimension $\frac{1}{2}n(n-1)$; that their ambient spaces, which we shall denote by L and M respectively, are each of dimension $\frac{1}{2}\binom{2n}{n} - 1 = \frac{1}{2}(N-1)$; and that these ambient spaces, which meet G only in G_α and G_β respectively, span the ambient space S_N of G.

We now prove

Proposition 1.1. *If P and P' are the image points of polar $[n-1]$'s Π and Π' with respect to Ω, then P and P' correspond in the harmonic homology of S_N with L and M as axial spaces.*

Proof. Let the equation of Ω, in a suitably chosen coordinate system of S_{2n-1}, be

$$z_0^2 + z_1^2 + \ldots + z_{2n-1}^2 = 0.$$

If Π is a generic $[n-1]$ of S_{2n-1}, join of the n points with coordinate vectors ξ_i $(i = 0, \ldots, n-1)$, then its polar space Π' is the intersection of the n primes with the same coordinate vectors ξ_i. This means, by the well-known relations between ordinary and dual Grassmann coordinates,‡ that if $p_0, \ldots, p_N$ are a suitably arranged set of coordinates for Π, then the coordinates $p'_0, \ldots, p'_N$ of Π' are equal to those of Π but interchanged by pairs, i.e.

$$p'_\alpha = p_\beta, \quad p'_\beta = p_\alpha,$$

where the $\frac{1}{2}(N+1)$ pairs (α, β) together make up the set $(0, 1, \ldots, N)$. Plainly then the image points P, P' of Π, Π' correspond in the harmonic homology of S_N with axial spaces L and M given respectively by

$$p_\alpha + p_\beta = 0 \quad \text{(all pairs } (\alpha, \beta))$$

and

$$p_\alpha - p_\beta = 0 \quad \text{(all pairs } (\alpha, \beta)).$$

Also L and M contain, evidently, the image varieties G_α and G_β, respectively, of the two systems of generators (self-polar $[n-1]$'s) of Ω. This proves the Proposition.

† For the proof in this case ($n = 4$), see Ch. 5, Prop. 4.1.

‡ See Hodge and Pedoe [**13**], Vol. 1, p. 294.

By a straightforward reversal of the above argument we have

COROLLARY 1.1.1. *If two points P, P' of G correspond in the harmonic homology of S_N with axial spaces L and M, then they represent spaces Π and Π' which are polars with respect to Ω.*

We now denote by σ the involution on G, with coincidence loci G_α and G_β, whose point-pairs represent pairs of polar spaces for Ω. Then two distinct points P, P' of G correspond in σ if and only if their join meets L and M in points U, V such that

$$\{P, P'; U, V\} = -1.$$

Further, if P is any point of G not lying on G_α or G_β, then the unique transversal line of L and M through P contains the point P' of G which corresponds in σ to P. If a line meets G in more than two points, then it lies entirely on G – as follows from the well-known result (Ch. 2, § 5) that any Grassmannian is an intersection of quadrics – and it then represents a pencil of $[n-1]$'s of S_{2n-1} (lying in an $[n]$ and passing through an $[n-2]$). Thus we may assert

COROLLARY 1.1.2. *The pairs (P, P') of the involution σ are in birational (though not unexceptional) correspondence with the chords of G which meet L and M.*

The exceptional elements in this correspondence arise from

(a) the points of G_α and G_β, since P coincides with P' at any such point so that the chord PP' is indeterminate, and

(b) the lines which lie on G and join a point of G_α to a point of G_β.

As regards (b), if p is one of the lines in question, then it represents a pencil of $[n-1]$'s, defined by a pair of generators α, β of Ω which meet in an $[n-2]$ and lie in an $[n]$; and then p contains an infinity of σ-pairs representing pairs of $[n-1]$'s of this pencil, polars for Ω, which separate α and β harmonically.

We now consider, as arising from Cor. 1.1.2, the correspondence between the σ-chords PP' – a system of dimension n^2 – and the points in which these chords meet one of the axial spaces, say L, which is of dimension $\frac{1}{2}\binom{2n}{n}-1$. Plainly, if $n = 2$ so that L is a plane, there must be infinitely many σ-chords through each point of L. We now assert however that if $n \geqslant 3$, so that

$$\frac{1}{2}\binom{2n}{n} - 1 \geqslant n^2,$$

then the σ-chords are in birational correspondence with the points in which they meet L. For any contrary hypothesis† would imply that there exist couples of σ-pairs, say (P, P') and (P_1, P_1'), representing pairs of skew polar $[n-1]$'s (Π, Π') and (Π_1, Π_1'), such that the σ-chords PP' and P_1P_1' meet on L; this again would imply that the four spaces Π, Π', Π_1, Π_1', being linearly dependent, have the property that every $[n-1]$ meeting three of them would also meet the fourth; and this, finally (cf. Ex. 1 below), can be shown to be impossible for $n > 2$ and under the conditions stated (Π skew to Π' and Π_1 skew to Π_1'). From this there follows

Corollary 1.1.3. *If $n \geqslant 3$, then the pairs (P, P') of the involution σ on G are in birational correspondence with the points of each of the two n^2-dimensional varieties which their joins PP' cut on L and M respectively.*

The two n^2-dimensional varieties mentioned in the above Corollary will be called *half-Grassmannians*.

By combining Cor. 1.1.2 and Cor. 1.1.3 we now obtain

Corollary 1.1.4. *If $n \geqslant 3$, then G projects doubly (i.e. by a $(2, 1)$-projection) from either of the axial spaces L or M into a half-Grassmannian variety Δ or Γ in the other. Each of these half-Grassmannians Γ, Δ becomes thereby a birational model of the polar pairs (Π, Π') of S_{2n-1} with respect to Ω.*

We now consider those curves on G – to be called C-curves – which represent Clifford reguli of S_{2n-1}; and we prove

Proposition 1.2. *The spaces of a regulus R of $[n-1]$'s of S_{2n-1} are represented on G by the points of a rational normal curve*

† The possible contrary hypotheses are that the projection of G from M onto a variety Γ in L is (*a*) dimension-lowering or (*b*) dimension-preserving but $(2N, 1)$ with $N > 1$; but both of these hypotheses lead easily to the implication as stated.

of order n; and R is a Clifford regulus if and only if this curve is compounded of the involution σ.

Proof. Without loss of generality we may suppose that the equation of R, in a suitably chosen coordinate system for S_{2n-1}, is $y = \lambda x$ (cf. Ch. 2, § 3). A simple calculation then shows that (apart from sign) the Grassmann coordinates of $y = \lambda x$ are just repetitions of $0, 1, \lambda, \ldots, \lambda^n$, and this proves the first assertion. The second follows at once from Cor. 1.1.1 using the autopolar property of a Clifford regulus.

If R is a Clifford regulus, then to the two terminal generators of R there will correspond two *terminal points* on the corresponding C-curve. Also, if n is even, these terminal points will either both lie on G_α or both on G_β; while, if n is odd, one of them will lie on G_α and the other on G_β. Hence:

COROLLARY 1.2.1. *If n is even, then the C-curves of one family on G meet G_α in two points and project doubly from L into curves of order $\frac{1}{2}(n-2)$ in M (not in general meeting G_β); while those of the second family (not meeting G_α) project doubly from L into curves of order $\frac{1}{2}n$ in M, each meeting G_β in two points. If n is odd, then the C-curves on G meet each of G_α and G_β in one point and project doubly from L into curves of order $\frac{1}{2}(n-1)$ in M, each meeting G_β in one point.*

We now note briefly, for reference, some general properties of the Grassmannian representation for the case of arbitrary $n \geqslant 3$. First, as regards the C-curves, and taking account of results already obtained in Ch. 3, Cor. 4.1.1, we note the following:

(i) The total freedom of C-curves on G, whether of two families for n even, or of one family for n odd, is $\frac{1}{2}(n-1)(3n+2)$.

(ii) The freedom of C-curves (of either family if n is even) through a general σ-pair (P, P') is $\frac{1}{2}n(n-1)$.

(iii) The freedom of C-curves through a general point of G_α or G_β is n^2-1.

(iv) The freedom of C-curves with given terminal points is $\frac{1}{2}(n-1)(n+2)$.

Next, as regards the half-Grassmannians Γ, Δ, in L, M respectively, we note:

(v) If P is any point of G_α, representing a generator α of Ω, then the generators β which each meet α in an $[n-2]$ are mapped on G_β by the points of a variety ψ_P which is the quadric Veronesean of an $[n-1]$.

(vi) In the (double) projection of G from L into Δ, the neighbourhood on G of any point P of G_α is projected into the ambient space – to be denoted by K_P – of the associated variety ψ_P of G_β; and the whole neighbourhood of G_α on G is projected accordingly into a primal ((n^2-1)-dimensional subvariety) of Δ – to be denoted by $\mathscr{K}$ – which is generated by the spaces K_P.

(vii) If P is any point of G_α, then the $[n]$'s that lie on G and pass through P fall into two systems; an $[n]$ of the first system represents the $[n-1]$'s of S_{2n-1} that lie in a fixed $[n]$ through the generator α corresponding to P, and an $[n]$ of the second system represents the $[n-1]$'s that pass through a fixed $[n-2]$ contained in α. Every line joining P to a point Q of ψ_P is the intersection of a unique pair of $[n]$'s of the two systems, and the two $[n]$'s of this pair project from L into the same tangent $[n-1]$ to ψ_P at Q.

We may note, finally, that if any C-curve passes through a point P of G_α, then its projection from L meets the space K_P which corresponds on Δ to the neighbourhood of P on G. Under this projection, in fact, the points of the space K_P represent (in general) the augmented generators based on the generator α represented by P.

Ex. 1. If $n \geqslant 3$ and if Π, Π', Π_1, Π_1' are four distinct $[n-1]$'s of S_{2n-1} such that Π is skew to Π' and Π_1 is skew to Π_1', prove that the four spaces cannot be such that every $[n-1]$ which meets three of them also meets the fourth.

[A proof along the following lines is suggested. First observe that, if Π_1' meets every $[n-1]$ which meets Π, Π' and Π_1, then it meets every *line* which meets these three spaces. Distinguish then between the cases (a) when all four spaces are mutually skew, and (b) when, say, Π meets Π_1 in a space S_h of dimension h. In case (a), by our first observation, the four spaces belong to a regulus; and this, by Prop. 1.2, is represented on G by a rational normal ${}^0C^n$ which contains no set of four coplanar points if

$n > 2$. In case (*b*), if $h \geqslant 1$, then the same observation tells us that Π_1' must meet every line joining a point of S_h to a point of Π'; and it must therefore contain S_h, in contradiction of the assumption that Π_1 is skew to Π_1'; and if, on the other hand, $h = 0$, then either Π_1' contains S_0 and the same argument applies; or Π_1' lies in the $[n]$ joining S_0 to Π', in which case Π_1' meets Π' in an $[n-2]$, and the previous argument applies to this intersecting pair.]

2. The cases $n = 2$ and $n = 3$

The case $n = 2$. This case requires only brief mention since it has already been dealt with in some detail (Ch. 4, § 1, Ex. 1).

We note, then, that G is here a quadric of S_5; and G_α, G_β are the conics in which G is met by a pair of skew polar planes L, M. Also σ is the involution of point-pairs of G whose joins meet L and M.

In this case, therefore, there is a birational correspondence between the point-pairs (P, P') of σ and the pairs of points (U, V) in which their joins PP' meet L and M respectively; and this is equivalent to a birational correspondence† between pairs of polar lines with respect to Ω and the point-pairs (U, V) of L and M. The exceptional elements arise from pairs (U, V) for which U lies on G_α or V on G_β, and also from pairs for which UV lies on G (i.e. U and V lie respectively on G_α and G_β).

The two types of Clifford reguli in S_3 (Ch. 4, § 1, Ex. 1) are represented on G by conics – sections of G by planes which meet one of L, M in a line and the other in a point. The two types of (maximal) linear ∞^2-systems in S_3 are represented by quadric surfaces – sections of G by solids through L and solids through M.

As already noted, the case $n = 2$ is exceptional; for only in this case does it happen that infinitely many of the chords PP' pass through any general point of L or M.

The case $n = 3$. Here Ω is a quadric V_4^2 of S_5; G is a variety V_9^{42} in S_{19}; the two systems of planes on Ω are mapped on G by varieties G_α and G_β which are quadric Veroneseans V_3^8 lying in skew 9-dimensional spaces L and M of S_{19}; and the involution σ on G is generated by the chords of G that meet L and M. In

† This is again, effectively, the Study correspondence, cf. Ch. 6, §3, Ex. 1.

this case (cf. Cor. 1.1.3 and Cor. 1.1.4), only one such chord passes through a general point of L or M, i.e. *L and M are themselves the half-Grassmannians Γ and Δ – double projections of G from M and L respectively*. Hence:

PROPOSITION 2.1. *When $n = 3$, G projects doubly from L onto M and from M onto L, each of the spaces L, M being therefore a birational model of the pairs of polar planes with respect to the quadric Ω in S_5.*

The (2, 1)-projection of G from L onto M has exceptional elements. In particular, the tangent [9] to G at a point P of G_α, since it meets L in the tangent solid to G_α at P, projects into a 5-dimensional space K_P of M; and all such spaces K_P generate a primal $\mathscr{K}$ of M representing the whole neighbourhood of G_α on G. Further there exist ∞^2 lines which lie on G and each join P to a point Q of G_β, and each of these lines contains ∞^1 pairs of mates in σ, separating P and Q harmonically. Further, by (v) of § 1, the ∞^2 points Q of G_β joined in this way to P form a Veronese surface ψ_P in G_β with K_P as its ambient space. Hence:

PROPOSITION 2.2. *For $n = 3$, each point P of G_α is joined by a cone of lines of G to a Veronese surface ψ_P of G_β (one of the ∞^3 such surfaces on G_β). The ambient space K_P of ψ_P is the projection from L of the tangent* [9] *to G at P; and the primal $\mathscr{K}$ generated in M by the ∞^3 spaces K_P represents by projection the whole neighbourhood of G_α on G.*

By Cor. 1.2.1, the Clifford reguli in S_5 – forming a single ∞^{11} system – are represented on G by twisted cubic curves (compounded of the involution σ) which meet each of G_α and G_β in one point; and each of them projects (doubly) from L into a line of M which meets G_β. Since already the lines of M which meet G_β form an irreducible ∞^{11} family, we have

PROPOSITION 2.3. *The ∞^{11} Clifford reguli of S_5 are in birational correspondence with the ∞^{11} lines of M which meet G_β.*

From the above we have a direct confirmation that there exist in S_5 no linear systems of Clifford parallel planes more ample than the Clifford regulus. For any such system would have to be represented in M by a variety (of dimension two or more) such that any general line joining two of its points lies on the variety and is unisecant to G_β; and no such variety exists because G_β, being a quadric Veronesean, possesses no linear subvarieties of positive dimension.

3. The case $n = 4$

Here Ω is a quadric Ω_6^2 of S_7; G is a 16-dimensional variety of order 24,024 in S_{69}; the two systems of solids on Ω are mapped on G by varieties G_α and G_β each of which (cf. Ch. 5, Prop. 4.1) is the quadratic Veronesean V_6^{128} of a 6-dimensional quadric; the two skew spaces L and M which contain G_α and G_β are each of dimension 34; and the involution σ on G is generated by the chords of G which meet L and M.

By Cor. 1.1.4, G projects doubly from L into a 16-dimensional variety Δ of M, and from M into a 16-dimensional variety Γ of L, each of the two half-Grassmannians Γ, Δ being a birational model of the pairs of polar solids with respect to Ω in S_7.

In the projection of G from L, the exceptional elements arise from (i) the points of G_α, and (ii) the lines which lie on G and join a point of G_α to a point of G_β. Considering these latter first, we observe (by (v) of §1) that the points Q of G_β that are joined to a given point P of G_α by lines lying on G form a threefold ψ_P, quadric Veronesean $V_3^8[9]$ of a solid. More precisely, if we denote by $\sqrt{G_\beta}$ the quadric of which G_β is the quadratic Veronesean, then ψ_P is a member of the ∞^6-family of quadric Veroneseans V_3^8 on G_β that correspond to the generating solids of one system on $\sqrt{G_\beta}$. Further, the ambient [9] of ψ_P is the projection K_P from L of the tangent [16] to G at P. (This [16] meets L in the tangent [6] to G_α at P.) Thus the whole neighbourhood of G_α on G projects from L into a primal (15-dimensional subvariety) of Δ generated by the ∞^6 9-dimensional spaces K_P; and this is the primal we have denoted by $\mathscr{K}$.

We now note further that through any point P of G_α there pass [4]'s which lie entirely on G and that these constitute two triply infinite families, say (λ) and (μ); for if α is the solid of Ω

represented by P, then the solids of S_7 which lie in a fixed [4] through α are mapped on a 4-space λ of G, and the solids of S_7 that meet α in a fixed plane are mapped on a 4-space μ of G. Again (from (vii) of § 1) we note that

(i) the λ and μ generate a cone, vertex P, whose section is a V_6^{20} – Segre product of two solids – and this cone is the model of all the solids of S_7 which meet α in a plane;

(ii) each λ is met by an associated μ in a line of the cone joining P to ψ_P; and

(iii) if PQ is any line of this cone, meeting G_β in Q, then the associated [4]'s λ and μ which meet in this line project from L into the same solid of K_P, namely the tangent solid to ψ_P at Q. Hence:

PROPOSITION 3.1. *For $n = 4$, the Grassmannian G projects doubly from L into the half-Grassmannian Δ, and the total neighbourhood of G_α on G projects into a primal $\mathscr{K}$ of Δ. The variety G_β contains ∞^6 quadric Veronesean threefolds ψ_P, each joined to a point P of G_α by a cone of lines of G. The tangent [16] to G at P projects from L into the ambient space K_P – a [9] – of ψ_P; and the ∞^6 spaces K_P generate the primal $\mathscr{K}$ of Δ. The two systems of [4]'s on G that pass through any point P of G_α project by pairs (one from each system) into the tangent solids of ψ_P.*

The C-curves on G, representing Clifford reguli of S_7, form two ∞^{21}-systems. By Cor. 1.2.1, they are all rational normal quartic curves, those of one system meeting G_α in two points and those of the other system meeting G_β in two points. Those of the first system project doubly from L into lines of Δ – chords of $\mathscr{K}$ – which we call the *α-lines* of Δ; while those of the second system project from L into conics of Δ – each meeting G_β in two points – which we call the *β-conics* of Δ. The same Clifford reguli are represented on the other half-Grassmannian Γ (projection of G from M) by the *α-conics* and *β-lines* of Γ respectively.

Consider now the two types of linear ∞^4-system of Clifford parallel solids in S_7 that we discussed in detail in Ch. 5, §§ 1 and 2, these two types being such that the terminal generators of their Clifford reguli are α-solids in the one case and β-solids in the other.

If $\mathscr{F}$ is a linear ∞^4-system of the first of these types, then it must be represented on Δ by a 4-fold F with the property that if (P, Q) is a generic point-pair of F, then the join of P to Q must be an α-line of Δ and this α-line must lie on F. Hence F is a 4-dimensional linear space lying on Δ. Further, since an α-line meets $\mathscr{K}$ in two points representing the terminal spaces of the corresponding Clifford regulus, it follows that F meets $\mathscr{K}$ in a quadric V_3^2 whose points represent the ∞^3 augmented α-solids that belong to $\mathscr{F}$. Finally, as may easily be verified, the quadric in question is non-singular and does not meet G_β. Recapitulating we have

PROPOSITION 3.2. *For $n = 4$, the two ∞^{21}-systems of Clifford reguli in S_7 are represented on Δ by α-lines, each meeting $\mathscr{K}$ in two points, and by β-conics, each meeting G_β in two points; and similarly on Γ they are represented respectively by α-conics, each meeting G_α in two points, and by β-lines. Further, a linear ∞^4-system of parallel solids of the first kind in S_7 (having α-solids as terminals of its Clifford reguli) is represented on Δ by a space S_4; and this S_4 meets $\mathscr{K}$ in a non-singular quadric threefold representing the augmented α-solids of the system. Similarly, a linear ∞^4-system of the second type in S_7 is represented on Γ by a space Σ_4, and this contains a non-singular quadric threefold representing the augmented β-solids of the system.*

Recalling now that there are ∞^{15} linear ∞^4-systems of each type, we remark that those of the first type, for example, correspond to ∞^{15} [4]'s lying on Δ, that each of these [4]'s contains ∞^6 α-lines, and that the ∞^{21} α-lines so arising account for the whole family of α-lines. This means, in effect, that a Clifford regulus in S_7 belongs to a finite number of linear ∞^4-systems of parallel solids; and in fact it belongs to two such systems (cf. Ch. 5, § 2, Ex. 3).

The remaining problem of this section is that of identifying the half-Grassmannians Γ and Δ (which are obviously projectively equivalent varieties) by finding a method of constructing them explicitly. What we find, in fact, is that, exceptionally for the case $n = 4$, each of Γ and Δ is projectively identical with the T-model for $n = 4$; and we prove this by showing that the previous construction of T (cf. Ch. 6, § 3) from its core variety $\mathscr{V}$, which for $n = 4$ is the quadratic Veronesean of a 6-dimensional quadric Ω', is reproduced exactly in a construction of Δ, for example, from its core variety – in this case G_β – which is likewise the quadratic Veronesean of a quadric $\sqrt{G_\beta}$.

Consider then Δ and its core variety G_β. What we have to show is that G_β uniquely determines the aggregate of α-lines of Δ, and therefore also Δ itself since the α-lines generate Δ multiply. As in the previous argument, we note first that if K_P and K_Q are the ambient [9]'s of quadric Veroneseans ψ_P and ψ_Q of G_β (each a $V_3^8[9]$), then there exists a symmetric determinantal Cremona transformation τ between K_P and K_Q, having ψ_P and ψ_Q as its key base varieties, which is such that a pair of points of K_P and K_Q correspond in τ if their join is an α-line. Further, as before, we note that any such transformation with ψ_P and ψ_Q as its key base varieties is uniquely determined by a correlation between ψ_P and ψ_Q – image of a correlation between the two skew generators of $\sqrt{G_\beta}$ of which ψ_P and ψ_Q are the Veronese transforms. We then observe that one such correlation between ψ_P and ψ_Q picks itself out from the others, namely that which corresponds to the natural correlation κ (cf. Ch. 2, § 2) between the two skew generators of $\sqrt{G_\beta}$; and we then see easily that it is precisely this correlation which determines the Cremona transformation τ between K_P and K_Q. Thus, as in the construction of T (Ch. 6, § 3), the core variety G_β of Δ determines the aggregate of α-lines of Δ, and hence Δ itself. This proves

Proposition 3.3. *For $n = 4$, each of the half-Grassmannians Γ and Δ is projectively equivalent to the corresponding T-model of the pairs of polar solids with respect to Ω in S_7.*

Although Γ, Δ and T are projectively equivalent to one another, the pairs of polar solids in S_7 are mapped on their points in very different ways; and in the following section we shall be considering the resulting (non-linear) birational correspondences between pairs of the three models in question. In preparation for this we need the following further property of the representation on Δ:

PROPOSITION 3.4. *There exists on Δ, in addition to its ∞^{21}-system of α-lines, a second ∞^{21}-system of lines – to be called the γ-lines of Δ – each of which represents a pair of pencils of solids in S_7 such that the solids of one pencil are polars of those of the other for Ω. Further these γ-lines (like the α-lines) lie in ∞^{15} [4]'s of a second system on Δ.*

Proof. It is clear, first of all, that any pencil of solids of S_7 is represented on G by a line l and that the polar pencil of solids will be represented on G by a second line $\sigma(l)$, image of l under the involution σ. Moreover, in the projection of G from L onto Δ, these two lines must project into the same line – a γ-line – of Δ.

Further, if π is any general plane of S_7, then the solids of S_7 through π are mapped on G by the points of a space S_4; and similarly, the solids of S_7 that lie in the polar [4] of π are mapped on G by the points of another space S'_4. Plainly then, in the projection from L, the two spaces S_4 and S'_4 project into the same [4] of Δ; and there are altogether ∞^{15} such [4]'s on Δ, corresponding to the ∞^{15} planes π of S_7. Each such [4] contains ∞^6 γ-lines representing polar pairs of pencils of S_7; and this proves the result.

COROLLARY 3.4.1. *There exists, similarly, a second system of lines on Γ (besides the β-lines); and, since these represent the same polar pairs of pencils of S_7 as the γ-lines of Δ, we shall call them the γ-lines of Γ.*

4. Construction and extension of the Study triality in S_7

For the following discussion it will be convenient, as regards the two types of Clifford regulus in S_7, to revert to a notation already introduced in Ch. 3, §2. A Clifford regulus whose terminal solids are both α-solids of Ω will therefore be called an *α-regulus;* and one whose terminals are β-solids of Ω will be called a *β-regulus.*

In the preceding section we obtained two birational mappings of the polar pairs (Π, Π') of solids of S_7 with respect to Ω, namely, that in which the (Π, Π') are mapped on the points of a half-Grassmannian Δ by the double projection of G from L into M, and that in which they are mapped on the points of Γ by the double projection of G from M into L.

As regards Δ, we saw that this model was completely determined by its core variety G_β, quadratic Veronesean of a 6-dimensional quadric which we denoted by $\sqrt{G_\beta}$. In particular:

(i) There exist on G_β two ∞^6-systems of quadric Veroneseans $V_3^8[9]$, to be called V_β and W_β respectively, which correspond to the two systems of generating solids of $\sqrt{G_\beta}$, to be denoted by $\sqrt{V_\beta}$ and $\sqrt{W_\beta}$ respectively. We take the V_β to be those Veroneseans which were previously called the ψ_P, recalling (from § 3) that each of them is joined to a point P of G_α by a cone of lines lying on G, and that it represents in S_7 the ∞^3 β-solids of Ω which meet a given α-solid in a plane. The W_β then each represent in S_7 the ∞^3 β-solids that pass through a fixed point of Ω.

(ii) There exist on Δ two primals (15-dimensional subvarieties), $\mathscr{K}$ and $\mathscr{J}$, such that $\mathscr{K}$ is generated by the ∞^6 ambient (9-dimensional) spaces K_β of the V_β, while $\mathscr{J}$ is generated by the ∞^6 ambient spaces J_β of the W_β. The points of $\mathscr{K}$ represent, in general, the augmented generators $\alpha(q)$ of Ω in S_7, those of them which lie in a space K_β representing the $\alpha(q)$ for which α is fixed (cf. § 3, where the notation K_P was used for a K_β).

(iii) There exist on Δ two ∞^{21}-systems of lines (each generating Δ multiply); and these are, respectively, the α-lines of Δ, each representing an α-regulus in S_7, and the γ-lines, each of which represents the polar pairs (Π, Π') of two mutually polar pencils of solids in S_7. The α-lines are the joins of corresponding points in Cremona transformations τ (determined by G_β) between every pair of the spaces K_β; and similarly the γ-lines are the joins of corresponding points in other Cremona transformations τ' (also determined by G_β) between pairs of the spaces J_β.

As regards Γ and its core variety G_α, similar properties hold. In particular, G_α has systems of Veroneseans V_α and W_α which correspond to the generating solids $\sqrt{V_\alpha}$ and $\sqrt{W_\alpha}$ of $\sqrt{G_\alpha}$; and each V_α represents the α-solids of Ω which meet a fixed β-solid in a plane, while each W_α represents the α-solids which pass through a fixed point of Ω. Also Γ possesses two ∞^{21}-systems of lines, namely its β-lines which represent β-reguli in S_7, and its γ-lines which each again represent the polar pairs (Π, Π') of two mutually polar pencils of solids of S_7.

We now concentrate our attention, in the first instance, on the relations between the three varieties Ω, G_α and G_β. As between Ω and G_β, we note the following correspondences:

β-solid of Ω	$\longleftrightarrow$	point of G_β,
α-solid of Ω (as defining the aggregate of β-solids which meet it in a plane)	$\longleftrightarrow$	V_β of G_β,
point of Ω (as defining the aggregate of β-solids which pass through it)	$\longleftrightarrow$	W_β of G_β.

More succinctly, if we replace G_β by the quadric $\sqrt{G_\beta}$ from which it is derived, the above correspondences take the form

$$\left.\begin{array}{lcl} \beta\text{-solid of } \Omega & \longleftrightarrow & \text{point of } \sqrt{G_\beta}, \\ \alpha\text{-solid of } \Omega & \longleftrightarrow & \text{solid } \sqrt{V_\beta} \text{ of } \sqrt{G_\beta}, \\ \text{point of } \Omega & \longleftrightarrow & \text{solid } \sqrt{W_\beta} \text{ of } \sqrt{G_\beta}. \end{array}\right\} \quad (4.1)$$

Similarly, as between Ω and $\sqrt{G_\alpha}$, we have the correspondences

$$\left.\begin{array}{lcl} \alpha\text{-solid of } \Omega & \longleftrightarrow & \text{point of } \sqrt{G_\alpha}, \\ \beta\text{-solid of } \Omega & \longleftrightarrow & \text{solid } \sqrt{V_\alpha} \text{ of } \sqrt{G_\alpha}, \\ \text{point of } \Omega & \longleftrightarrow & \text{solid } \sqrt{W_\alpha} \text{ of } \sqrt{G_\alpha}. \end{array}\right\} \quad (4.2)$$

Finally, as between G_α and G_β, we make two points correspond if their join lies on G; and this induces a correspondence between $\sqrt{G_\alpha}$ and $\sqrt{G_\beta}$ in which

$$\left.\begin{array}{lcl} \text{point of } \sqrt{G_\alpha} & \longleftrightarrow & \text{solid } \sqrt{V_\beta} \text{ of } \sqrt{G_\beta}, \\ \text{solid } \sqrt{V_\alpha} \text{ of } \sqrt{G_\alpha} & \longleftrightarrow & \text{point of } \sqrt{G_\beta}, \\ \text{solid } \sqrt{W_\alpha} \text{ of } \sqrt{G_\alpha} & \longleftrightarrow & \text{solid } \sqrt{W_\beta} \text{ of } \sqrt{G_\beta}, \end{array}\right\} \quad (4.3)$$

where, in this last entry, we mean that a solid $\sqrt{W_\alpha}$ of $\sqrt{G_\alpha}$ corresponds to a solid $\sqrt{W_\beta}$ of $\sqrt{G_\beta}$ if they represent the same point of Ω in the correspondences (4.1) and (4.2).

We have thus derived geometrically the system consisting of three 6-dimensional quadrics, in mutual correspondence, which is known as the Study triality.†

† There is an extensive literature dealing with the Study triality, and the reader interested in this topic may consult the bibliographies of two recent papers on the subject: Tits[**25**] and Dye[**8**].

The principal property of the triality is that each of the three correspondences preserves *incidence*, where incidence is defined as follows:

(*a*) a point is incident with a solid if it lies in the solid;

(*b*) two points of a (6-dimensional) quadric are incident if they are conjugate points, i.e. if their join lies on the quadric;

(*c*) two solids of the same system on a quadric are incident if they meet in a line; and

(*d*) two solids of opposite systems on a quadric are incident if they meet in a plane.

The proof that all these incidence relations are preserved under the triality – and other known results besides – follow very easily in terms of our model, and we leave the reader to fill in the details.

We now recall the interesting coincidence (Prop. 3.3) that in the present case ($n = 4$) the T-model is projectively the same variety as each of the half-Grassmannians Γ and Δ. If $\mathscr{V}$ denotes the core variety of T, then $\mathscr{V}$ is the quadratic Veronesean of a 6-dimensional quadric that we originally denoted by Ω' (Ch. 6, § 3) but which we may now denote for consistency by $\surd\mathscr{V}$. We also denote the two systems of Veroneseans $V_3^8[9]$ on $\mathscr{V}$ by ϕ and ψ, remembering that these correspond to generating solids of $\surd\mathscr{V}$ which we denote by $\surd\phi$ and $\surd\psi$. We recall, then, that the ambient spaces A of the Veroneseans ϕ generate a primal $\mathscr{A}$ of T, whose points represent in general the augmented generators $\alpha(q)$ of Ω in S_7, and, similarly, the ambient spaces B of the ψ generate a primal $\mathscr{B}$ of T whose points represent in general the augmented generators $\beta(q)$. Also T contains two ∞^{21}-systems of lines, its α-lines representing α-reguli and its β-lines representing β-reguli, such that each of the former meets a pair of the spaces A in points which correspond in a Cremona transformation (determined by $\mathscr{V}$) between the spaces; while the β-lines are similarly defined.

We now make points of Γ, Δ and T correspond to each other if they represent the same polar pair in S_7, and we thereby obtain three commuting birational correspondences λ, μ, ν as indicated in the diagram:

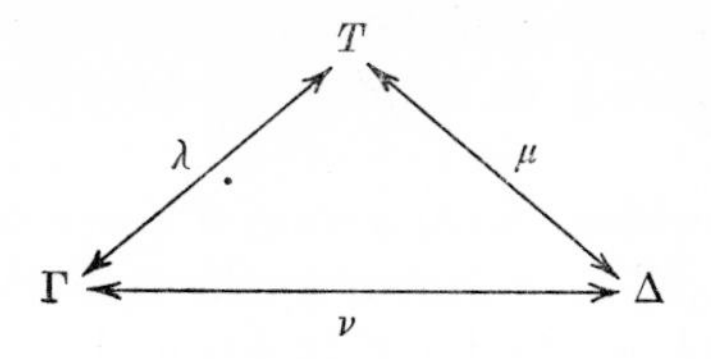

These correspondences, of course, are not trivial (i.e. not collineations); for an α-regulus is represented on each of T and Δ by a line, but on Γ by a conic; whereas a β-regulus is represented on each of T and Γ by a line, but on Δ by a conic. Further, it is easy to verify that a system of polar pairs of the kind that is represented on Γ or Δ by a γ-line is represented on T by a conic. Hence:

PROPOSITION 4.1. *There exist on each of the (projectively equivalent) varieties Γ, Δ and T three ∞^{21}-systems of curves (either lines or conics), representing α-reguli, β-reguli and pairs of mutually polar pencils of solids, respectively, in S_7. The birational correspondences (induced by the representations) between pairs of Γ, Δ and T transform the triads of curve systems in question according to the columns of the table.*

	α-regulus	β-regulus	Pair of mutually polar pencils of solids
Γ	α-conic	β-line	γ-line
Δ	α-line	β-conic	γ-line
T	α-line	β-line	γ-conic

As regards the three birational correspondences λ, μ, ν, it is clear that each of the three core varieties, G_α, G_β and $\mathscr{V}$, is fundamental for those two of λ, μ, ν that carry the corresponding variety Γ, Δ or T into either of the other two, transforming it in each case into a primal of the other. We note, in fact

COROLLARY 4.1.1. *The restriction of each of the birational correspondences λ, μ, ν to the core varieties is an ∞^9-correspondence, and these ∞^9-correspondences induce, between the quadrics $\sqrt{G_\alpha}$, $\sqrt{G_\beta}$ and $\sqrt{\mathscr{V}}$, the three correspondences of a Study triality.*

APPENDIX

THE HURWITZ–RADON MATRIX EQUATIONS

As we have shown in Ch. 4, the existence of an r-dimensional non-singular linear system of Clifford parallels in S_{2n-1} ($n \geqslant 2$) is equivalent, algebraically, to the existence of a set of $r-1$ (complex) matrices $S_1, \ldots, S_{r-1}$ of order n satisfying the Hurwitz–Radon matrix equations

$$\left.\begin{aligned} S_i^T &= -S_i, \quad S_i^2 = -I \quad (i = 1, \ldots, r-1), \\ S_i S_j &= -S_j S_i \quad (i,j = 1, \ldots, r-1;\, i \neq j). \end{aligned}\right\} \tag{0.1}$$

The object of this appendix is to establish the following

THEOREM A 1. *If n is written in the form $n = u\,.\,2^{4\alpha+\beta}$ (with u odd and $0 \leqslant \beta \leqslant 3$) and if $H(n) = 8\alpha + 2^\beta$, then the equations* (0.1) *admit solutions in matrices of order n if and only if $r \leqslant H(n)$. In particular, solutions of* (0.1), *with $r = n$, exist only in the three cases $n = 2$, 4 and* 8.

It will appear, incidentally, that there are, up to orthogonal equivalence, only a finite number of solution sets of (0.1) for given n and r; so that, by Ch. 4, Cor. 3.1.2, there are only a finite number of projectively distinct non-singular linear systems of Clifford parallels in S_{2n-1}.

The proof of Theorem A 1 will be given in § 2 below. Before this, however, we include some brief historical notes by way of introduction.

1. Historical notes

The equations (0.1) appeared first in the literature in connection with the problem of composition of quadratic forms, which may be stated as follows.

Find the values of n and r for which there exist n forms $z_1, \ldots, z_n$, bilinear with complex coefficients in the two sets of indeterminates $x_1, \ldots, x_r$ and $y_1, \ldots, y_n$, satisfying the identity

$$(x_1^2 + \ldots + x_r^2)(y_1^2 + \ldots + y_n^2) \equiv z_1^2 + \ldots + z_n^2. \tag{1.1}$$

This problem, which generalizes the famous eight squares problem, was first proposed and solved by Hurwitz in his posthumous paper [**15**]. The same author had solved the problem in the case $n = r$ in an earlier paper [**14**].

To see how this problem reduces to the equations (0.1), suppose that

$$z_i = \sum_{j=1}^{n} \sum_{h=1}^{r} a_{ihj} x_h y_j \quad (i = 1, \ldots, n), \tag{1.2}$$

and, for $h = 1, \ldots, r$, let A_h denote the matrix of order n whose (i,j) element is a_{ihj}. If, then, we write

$$\Lambda = x_1 A_1 + \ldots + x_r A_r,$$

we see that the (i,j) element of Λ is

$$\lambda_{ij} = \sum_{h=1}^{r} x_h a_{ihj}$$

and (1.2) may be written

$$z_i = \sum_{j=1}^{n} \lambda_{ij} y_j \quad (i = 1, \ldots, n).$$

Hence, using x, y, z to denote the column vectors (x_i), (y_i), (z_i), we have $z = \Lambda y$ and the desired identity (1.1) becomes

$$(x^T x)(y^T y) \equiv z^T z \equiv y^T \Lambda^T \Lambda y;$$

this is plainly possible if and only if $\Lambda^T \Lambda \equiv (x^T x) I$ or, in full, if and only if

$$(x_1 A_1^T + \ldots + x_r A_r^T)(x_1 A_1 + \ldots + x_r A_r) \equiv (x_1^2 + \ldots + x_r^2) I.$$

Thus, comparing coefficients on either side of this identity, we require

$$A_i^T A_i = I \quad (i = 1, \ldots, r)$$

and

$$A_i^T A_j = -A_j^T A_i \quad (i,j = 1, \ldots, r;\ i \neq j).$$

Finally, writing $A_i = A_r S_i$ $(i = 1, \ldots, r-1)$, these last equations reduce precisely to the form (0.1), together with $A_r^T A_r = I$.

Contemporaneously with Hurwitz, Radon [**20**] showed that the inequality of Theorem A 1 is necessary and sufficient for the existence of *real* solutions of (0.1). This means, of course, that whenever equations (0.1) admit complex solutions, then they necessarily also admit real solutions; this fact was made particularly clear in a subsequent proof of Theorem A 1 by Eckmann [**9**]. Another proof of Radon's result was given by Lee [**18**]. In another direction, Albert [**2**] and Dubisch [**7**] solved the composition problem for quadratic forms over more general fields, and later authors have generalized and extended their work. Finally, with a slightly different emphasis, Wong [**27**] discussed the equations (0.1) with particular reference to the *maximal real* solutions (i.e. the solutions of (0.1) that cannot be extended† to solutions involving more than $r-1$ matrices).

No account of the equations (0.1) would be complete without mention of their close connection with a well-known problem concerning vector fields. We recall that a *q-field* on a smooth manifold M is a set of q smooth tangent vector fields on M such that, at every point of M, the vectors belonging to the q fields are linearly independent. An obviously important question is that of determining the maximum possible q for a given manifold M.

Let us say that a set of p real matrices $A_1, \ldots, A_p$ of order n is a *strongly linearly independent set* if, for all non-zero real vectors x, the vectors

$$A_1 x, \ldots, A_p x$$

are linearly independent. (This is equivalent to saying that 0 is the only singular matrix in the real linear family

$$\lambda_1 A_1 + \ldots + \lambda_p A_p.)$$

Now Stiefel [**24**] proved that, if there exists a strongly linearly independent set of p matrices of order n, then there exists a $(p-1)$-field on the real projective $(n-1)$-space, and consequently also on the $(n-1)$-sphere. However, if $S_1, \ldots, S_{r-1}$ is a set of real matrices satisfying (0.1) and if $B = \alpha I + \lambda_1 S_1 + \ldots + \lambda_{r-1} S_{r-1}$, then

$$B^T B = (\alpha^2 + \lambda_1^2 + \ldots + \lambda_{r-1}^2) I,$$

from which it follows that $I, S_1, \ldots, S_{r-1}$ is a strongly linearly independent set. We deduce that, *if the integers r and n are such that the equations* (0.1) *admit solutions, then there exists an* $(r-1)$*-field on real projective* $(n-1)$*-space and on the* $(n-1)$*-sphere.*

† Cf. Ch. 4, §5, Ex. 2.

More recently, Adams [**1**] has solved the converse problem of determining all cases in which an $(r-1)$-field may exist on an $(n-1)$-sphere. Using powerful techniques of K-theory, he proves that *such fields can exist if and only if the inequality of Theorem* A 1 *is satisfied*. A brief account of Adams' work, and of a number of other related questions, is given by Atiyah [**3**]. The reader may also consult Husemoller [**16**], particularly the chapter on Clifford algebras.

2. Eckmann's proof of Theorem A1

The proof to be given here of Theorem A 1 follows closely the account given by Eckmann [**9**], which was inspired by a proof for the special case $n = r$ given by Jordan, von Neumann and Wigner [**17**]. In addition to being quite short, it provides an elegant and simple example of the application of general theorems from the representation theory of finite groups (in contrast to the proofs of Hurwitz and Radon, which employ largely *ad hoc* methods, outside the mainstream of modern algebra). The results needed from group representation theory are elementary and may be found in Serre [**22**].

We observe, first of all, that a skew-symmetric matrix S_i satisfies $S_i^2 = -I$ if and only if it is (complex) orthogonal. We are concerned, then, with the problem of determining values of r and n for which there exist orthogonal matrices $S_1, \ldots, S_{r-1}$ of order n satisfying

$$\left.\begin{aligned} S_i^2 &= -I \quad (i = 1, \ldots, r-1), \\ S_i S_j &= -S_j S_i \quad (i,j = 1, \ldots, r-1;\ i \neq j). \end{aligned}\right\} \tag{2.1}$$

Let G denote the abstract group on r generators $e, s_1, \ldots, s_{r-1}$ subject to the relations†

$$\left.\begin{aligned} &e^2 = 1, \\ &s_i^2 = e, \quad s_i e = e s_i \quad (i = 1, \ldots, r-1), \\ &s_i s_j = e s_j s_i \quad (i,j = 1, \ldots, r-1;\ i \neq j). \end{aligned}\right\} \tag{2.2}$$

† We present the relations for G in the form given by Eckmann [**9**]; but it may be noted that the relations $s_i e = e s_i$ can be omitted (being consequences of $s_i^2 = e$).

Our problem, as just stated, is plainly equivalent to that of determining whether there exists a representation of G by (complex) orthogonal matrices of order n, with e represented by the matrix $-I$. To solve the problem in this form we shall regard r as fixed and proceed in four steps:

(A) we first find the degrees of the *irreducible* representations of G in which e is represented by $-I$ (ignoring temporarily the requirement of orthogonality);

(B) we next find the degrees of *all* representations of G in which e is represented by $-I$;

(C) we then decide which of the representations found in (B) are equivalent to orthogonal representations; and finally

(D) we find that an orthogonal representation, in which e is represented by $-I$, can be of degree n if and only if the inequality of Theorem A 1 is satisfied.

(A) To begin with, it follows from the relations (2.2) that all the elements of G are of the form

$$s_{i_1} s_{i_2} \dots s_{i_k} \quad \text{or} \quad e s_{i_1} s_{i_2} \dots s_{i_k}, \tag{2.3}$$

where $0 \leqslant k \leqslant r-1$ and the integers $i_1, i_2, \dots, i_k$ take all possible sequences of values from $1, \dots, r-1$ such that $i_1 < i_2 < \dots < i_k$. The order of G is therefore 2^r, and an explicit list of its elements is

$$\left.\begin{array}{l} 1;\, s_1, \dots, s_{r-1};\, s_1 s_2, \dots, s_{r-2} s_{r-1};\, \dots;\, s_1 s_2 \dots s_{r-1}; \\ e;\, e s_1, \dots, e s_{r-1};\, e s_1 s_2, \dots, e s_{r-2} s_{r-1};\, \dots;\, e s_1 s_2 \dots s_{r-1}. \end{array}\right\} \tag{2.4}$$

When $r = 2$, G is the cyclic group of order four and this, as is well-known, possesses just four irreducible representations (each of degree one). *We assume from now on that* $r \geqslant 3$.

It is easy to verify that each commutator of G is either 1 or e; so the commutator subgroup K of G is of order two and the abelian group $G' = G/K$ is of order 2^{r-1}. Hence G *possesses* 2^{r-1} *inequivalent representations of degree one.*

It is again easy to verify that a complete set of conjugates of an element g of G consists either of g alone or of g and eg. More precisely:

(*a*) If r is odd, then 1 and e are conjugate only to themselves, while every other element g is conjugate to eg.

(*b*) If r is even, then $1, e, s_1 s_2 \ldots s_{r-1}$ and $es_1 s_2 \ldots s_{r-1}$ are conjugate to themselves, while every other element g is conjugate to eg.

Thus, letting h denote the number of conjugacy classes in G, we have:

(*a*) If r is odd, there are two conjugacy classes with one element and $\frac{1}{2}(2^r - 2)$ classes with two elements, so that $h = 2^{r-1} + 1$.

(*b*) If r is even, there are four conjugacy classes with one element and $\frac{1}{2}(2^r - 4)$ classes with two elements, so that $h = 2^{r-1} + 2$.

Since the number of inequivalent irreducible representations of G is h, we have (besides the 2^{r-1} representations of degree one noted earlier)

(*a*) if r is odd, just one irreducible representation of degree $d > 1$,

(*b*) if r is even, just two irreducible representations of degrees d, d' $(d > 1, d' > 1)$.

Moreover, since the order of a group is equal to the sum of the squares of the degrees of its irreducible representations, the values of d and d' are easily calculated; namely

(*a*) if r is odd, then $d^2 + 2^{r-1} = 2^r$, so that $d = 2^{\frac{1}{2}(r-1)}$;

(*b*) if r is even, then $d^2 + d'^2 + 2^{r-1} = 2^r$, so that

$$d^2 + d'^2 = 2^{r-1};$$

and, remembering that d and d' must divide the order 2^r of G, we must have $d = 2^c$, $d' = 2^{c'}$ and hence

$$2^{2c} + 2^{2c'} = 2^{r-1},$$

from which it follows that $c = c' = \frac{1}{2}(r-2)$ and $d = d' = 2^{\frac{1}{2}(r-2)}$.

In the 2^{r-1} representations of G of degree one, e is represented by the unit matrix $I = (1)$. On the other hand, since the group $G' = G/(1, e)$ is abelian, e cannot be represented by the unit matrix in the irreducible representations of higher degree. But, since e lies in the centre of G, its corresponding matrix (which must commute with all the matrices of the representation) is necessarily a multiple λI of the unit matrix; and, since $e^2 = 1$, we must have $\lambda^2 = 1$, so that $\lambda = -1$ and e is represented by $-I$. Hence:

There is, for r odd, just one irreducible representation of G (up to equivalence) which satisfies our condition that e be represented *by* $-I$, *and it is of degree* $2^{\frac{1}{2}(r-1)}$; *while, for* n *even, there are two such representations, each of degree* $2^{\frac{1}{2}(r-2)}$. *The latter statement is valid also when* $r = 2$, *a case we excluded earlier.*

(B) It follows now that the degree n of any (not necessarily irreducible) representation of G, in which e is represented by $-I$, is of the form

(a) $n = m \, . \, 2^{\frac{1}{2}(r-1)}$ if r is odd,

(b) $n = m \, . \, 2^{\frac{1}{2}(r-2)}$ if r is even,

where m is a positive integer. In case (a) there is, up to equivalence, a unique representation (of the required kind) of degree n; whereas, in case (b), there are $m+1$ inequivalent representations of degree n. The two cases may be conveniently summed up in the following statement:

If $n = u \, . \, 2^t$, *where* u *is odd, then there exists a representation of* G *of degree* n, *with* e *represented by* $-I$, *if and only if* $r \leqslant 2t+2$.

(C) We now proceed to pick out, from among the representations of G that we have found, those which are equivalent to orthogonal representations. These, by a result of Frobenius and Schur, are precisely those which are equivalent to real representations (and hence also to real orthogonal representations). First, then, we must discover, from among the *irreducible* representations that we have found, those which are equivalent to real representations.

To this end, we consider the sum

$$F = \sum_{g \in G} \chi(g^2)$$

of the characters† of the squares of all the elements $g \in G$, recalling that an irreducible representation D is (i) equivalent to a real one if and only if the character-sum F is positive, (ii) not equivalent to a real representation, but equivalent to its complex conjugate representation $\bar{D}$, if F is negative, and (iii) not equivalent to a real representation, nor to $\bar{D}$, if $F = 0$.

Now (cf. equation (2.3)), any element $g \in G$ is of the form

$$g = s_{i_1} s_{i_2} \ldots s_{i_k} \quad \text{or} \quad g = e s_{i_1} s_{i_2} \ldots s_{i_k},$$

† The character of an element $g \in G$, in a given representation of G, is the trace of the matrix corresponding to g.

and in either case we find easily, by use of (2.2), that

$$g^2 = e^{\frac{1}{2}k(k+1)},$$

so that $\qquad g^2 = 1 \quad \text{if} \quad k \equiv 3 \text{ or } 0 \pmod 4$

and $\qquad g^2 = e \quad \text{if} \quad k \equiv 1 \text{ or } 2 \pmod 4.$

Hence, if a representation D of G carries e into $-I$ and is of degree d, we have

$$\chi(g^2) = d \qquad \text{if} \quad k \equiv 3 \text{ or } 0 \pmod 4$$

and $\qquad \chi(g^2) = -d \quad \text{if} \quad k \equiv 1 \text{ or } 2 \pmod 4.$

Thus, summing over the 2^r elements of G given by (2.4), we have

$$F = \sum_{g \in G} \chi(g^2) = 2d\sigma,$$

where σ denotes the sum

$$\sigma = \binom{r-1}{0} - \binom{r-1}{1} - \binom{r-1}{2} + \binom{r-1}{3} + \binom{r-1}{4} - \ldots \pm \binom{r-1}{r-1}.$$

To determine the sign of σ, we consider the complex number

$$z = (1-i)^{r-1} = x + iy \quad (x, y \text{ real}),$$

noting that σ is just $x+y$. Thus, since $\arg z = -\frac{1}{4}\pi(r-1)$, we have the following cases:

$$\sigma > 0 \quad \text{if} \quad -\frac{\pi}{4}(r-1) \equiv 0, \quad \frac{\pi}{4} \quad \text{or} \quad \frac{2\pi}{4} \pmod{2\pi},$$

$$\sigma = 0 \quad \text{if} \quad -\frac{\pi}{4}(r-1) \equiv \frac{3\pi}{4} \quad \text{or} \quad \frac{7\pi}{4} \pmod{2\pi},$$

$$\sigma < 0 \quad \text{if} \quad -\frac{\pi}{4}(r-1) \equiv \frac{4\pi}{4}, \quad \frac{5\pi}{4} \quad \text{or} \quad \frac{6\pi}{4} \pmod{2\pi}.$$

It follows that F is positive if $r \equiv 7$, 0 or 1 (mod 8), negative if $r \equiv 3$, 4 or 5 (mod 8), and zero if $r \equiv 2$ or 6 (mod 8). Hence:

The irreducible representations of G found earlier are equivalent to real representations when $r \equiv 7$, 0 or 1 (mod 8), but not for other values of r. When $r \equiv 3$, 4 or 5 (mod 8), they are equivalent to their complex conjugate representations, but not so when $r \equiv 2$ or 6 (*mod* 8). *When* $r \equiv 2$ *or* 6 (*mod* 8), *each of the two irreducible representations is equivalent to the complex conjugate of the other.*

It follows now that the degree n of a real representation of G, in which e is represented by $-I$, must be an even multiple of the degree of the irreducible representation(s),† except when $r \equiv 7$, 0 or 1 (mod 8). The degree of a real-irreducible representation must therefore be equal to

$$\left.\begin{aligned}
&(a)\ 2^{\frac{1}{2}(r-1)} \quad \text{if} \quad r \equiv 7 \text{ or } 1 \pmod 8,\\
&(b)\ 2^{\frac{1}{2}(r-2)} \quad \text{if} \quad r \equiv 0 \pmod 8,\\
&(c)\ 2.2^{\frac{1}{2}(r-1)} = 2^{\frac{1}{2}(r+1)} \quad \text{if} \quad r \equiv 3 \text{ or } 5 \pmod 8,\\
&(d)\ 2.2^{\frac{1}{2}(r-2)} = 2^{\frac{1}{2}r} \quad \text{if} \quad r \equiv 2, 4 \text{ or } 6 \pmod 8.
\end{aligned}\right\} \tag{2.5}$$

Moreover there are, up to equivalence, two such representations when $r \equiv 0$ or 4 (mod 8), while there is only one for other values of r. Hence:

Orthogonal representations of G of degree $n = m.2^q$ (where m is a positive integer), in which e is represented by $-I$, exist only in the following cases:

$$\left.\begin{aligned}
&(a)\ q \equiv 3 \text{ or } 0 \pmod 4 \quad \textit{and} \quad r = 2q+1,\\
&(b)\ q \equiv 3 \pmod 4 \quad \textit{and} \quad r = 2q+2,\\
&(c)\ q \equiv 2 \text{ or } 3 \pmod 4 \quad \textit{and} \quad r = 2q-1,\\
&(d)\ q \equiv 1, 2 \text{ or } 3 \pmod 4 \quad \textit{and} \quad r = 2q.
\end{aligned}\right\} \tag{2.6}$$

(D) So far, we have been thinking of r as fixed and have found all possible values of n for which the required representations exist. If we now reverse this emphasis, thinking of n as given and looking for the possible values of r, we see easily that (2.6) is equivalent to the following:

If $n = u.2^t$ (with u odd), then the largest possible value $H(n)$ of r is given by

$$\begin{aligned}
H(n) &= 8\alpha+1 \quad \textit{when} \quad t = 4\alpha,\\
H(n) &= 8\alpha+2 \quad \textit{when} \quad t = 4\alpha+1,\\
H(n) &= 8\alpha+4 \quad \textit{when} \quad t = 4\alpha+2,\\
H(n) &= 8\alpha+8 \quad \textit{when} \quad t = 4\alpha+3.
\end{aligned}$$

† A real representation of least degree, for $r = 2, \ldots, 6$ (mod 8), is equivalent to the direct sum of an irreducible representation with its conjugate complex representation.

Combining these four possibilities into one, namely,

$$H(n) = 8\alpha + 2^\beta \quad \text{when} \quad t = 4\alpha + \beta \quad (0 \leqslant \beta \leqslant 3),$$

we have finally the result of Theorem A 1.

If r satisfies the condition $r \leqslant H(n)$ then, by the note following equations (2.5) above, there is exactly one solution (up to equivalence) of the equations (0.1) when $r \not\equiv 0 \pmod 4$; whereas, if $r \equiv 0 \pmod 4$, there are $m+1$ different solutions, where

$$m = \frac{n}{2^{\frac{1}{2}(r-2)}} \quad \text{when} \quad r \equiv 0 \pmod 8,$$

and

$$m = \frac{n}{2^{\frac{1}{2}r}} \quad \text{when} \quad r \equiv 4 \pmod 8.$$

The explicit solutions, in all cases, may easily be deduced from the forms given by Hurwitz [**15**], although this author takes the original equations (0.1) in a slightly different form.

Finally, we again draw attention to the fact (already noted in Ch. 4, § 5, Ex. 2) that a solution of (0.1), with $r < H(n)$, is not necessarily extendable to a solution with $r = H(n)$.

3. Explicit solutions for the cases n = 2, 4 and 8

For completeness, we give here the solutions (up to orthogonal equivalence) of the equations (0.1) in the three cases of most interest, $n = 2$, 4 and 8.

We first define matrices I, J, K, L of order two as follows:

$$I = \begin{pmatrix} 1 & 0 \\ 0 & 1 \end{pmatrix}, \quad J = \begin{pmatrix} 0 & -1 \\ 1 & 0 \end{pmatrix}, \quad K = \begin{pmatrix} 1 & 0 \\ 0 & -1 \end{pmatrix}, \quad L = \begin{pmatrix} 0 & 1 \\ 1 & 0 \end{pmatrix},$$

noting that

$$J^2 = -I, \quad K^2 = L^2 = I,$$

$$KL = -LK = -J,$$

$$LJ = -JL = K,$$

$$JK = -KJ = L.$$

(i) If $n = 2$, then $H(n) = 8\alpha + 2^\beta = 2$ and the only solutions of $S_1^2 = -I$ are $S_1 = J$ and $S_1 = -J$. The two solutions are orthogonally equivalent.

(ii) If $n = 4$, then $H(n) = 4$. One solution of (0.1) with $r = 4$ is†

$$S'_1 = I \times J, \quad S'_2 = J \times K, \quad S'_3 = J \times L$$

and every solution with $r = 4$ is orthogonally equivalent either to (S'_1, S'_2, S'_3) or to $(-S'_1, -S'_2, -S'_3)$. The solution of (0.1) with $r = 3$ (resp. $r = 2$) is unique up to equivalence, an explicit solution being (S'_1, S'_2) (resp. (S'_1)).

(iii) If $n = 8$, then $H(n) = 8$. One solution of (0.1) with $r = 8$ is

$$\begin{aligned} S''_1 &= I \times I \times J, & S''_2 &= I \times J \times K, \\ S''_3 &= J \times I \times L, & S''_4 &= J \times K \times K, \\ S''_5 &= J \times L \times K, & S''_6 &= K \times J \times L, \\ S''_7 &= L \times J \times L, \end{aligned}$$

and every solution with $r = 8$ is orthogonally equivalent either to $(S''_1, \ldots, S''_7)$ or to $(-S''_1, \ldots, -S''_7)$. If $r < 8$ and $r \neq 4$, the solution of (0.1) is unique up to equivalence, an explicit solution being $(S''_1, \ldots, S''_{r-1})$. If $r = 4$, however, there are three inequivalent solutions, one of which may be taken to be (S''_1, S''_2, S''_3); the other two are, modulo equivalence,

$$I \times I \times J, \quad I \times J \times K, \quad I \times J \times L$$

and its negative. These two 'extra' solutions are not extendable to solutions involving more than three matrices; the geometrical significance of this has already be explained (Ch. 5, § 4).

† The cross denotes Kronecker product of matrices, and the solutions may be easily checked by use of the basic multiplication rule

$$(A \times B)(C \times D) = AC \times BD.$$

REFERENCES

[**1**] Adams, J. F. Vector fields on spheres. *Annals of Math.* (2) **75** (1962), 603–632.

[**2**] Albert, A. A. Quadratic forms permitting composition. *Annals of Math.* (2) **43** (1942), 161–177.

[**3**] Atiyah, M. F. The role of algebraic topology in mathematics. *J. London Math. Soc.* **41** (1966), 63–69.

[**4**] Burau, W. *Mehrdimensionale projektive und höhere Geometrie* (Berlin, 1961).

[**5**] Clifford, W. K. Preliminary sketch of biquaternions. *Proc. London Math. Soc.* (1) **4** (1873), 381–395. Reprinted in *Mathematical Papers*, 181–200.

[**6**] Coxeter, H. S. M. *Non-Euclidean geometry* (Toronto, 1942).

[**7**] Dubisch, R. Composition of quadratic forms. *Annals of Math.* (2) **47** (1946), 510–527.

[**8**] Dye, R. H. The simple group FH(8, 2) of order $2^{12}.3^{5}.5^{2}.7$ and the associated geometry of triality. *Proc. London Math. Soc.* (3) **18** (1968), 521–562.

[**9**] Eckmann, B. Gruppentheoretischer Beweis des Satzes von Hurwitz–Radon über die Komposition quadratischer Formen. *Comment. Math. Helv.* **15** (1943), 358–366.

[**10**] Forsyth, A. R. *Geometry of four dimensions*, Vol. 1 (Cambridge University Press, 1930), particularly Chapter VI.

[**11**] Heymans, P. Pfaffians and skew-symmetric matrices. *Proc. London Math. Soc.* (3) **19** (1969), 730–768.

[**12**] Heymans, P. The minimum model of spaces on a quadric (to appear in *Proc. London Math. Soc.*).

[**13**] Hodge, W. V. D. and Pedoe, D. *Methods of algebraic geometry*, Vols. 1 and 2 (Cambridge University Press, 1947 and 1952).

[**14**] Hurwitz, A. Über die Komposition der quadratischen Formen von beliebig vielen Variablen. *Nach. v. der Ges. der Wiss. Göttingen, Math.-Phys. Klasse* (1898), 309–316. Reprinted in *Math. Werke, Bd. 2*, 565–571.

[**15**] Hurwitz, A. Über die Komposition der quadratischen Formen. *Math. Ann.* **88** (1923), 1–25. Reprinted in *Math. Werke, Bd. 2*, 641–666.

[**16**] Husemoller, D. *Fibre bundles* (McGraw-Hill, New York, 1966).

[**17**] Jordan, P., Von Neumann, J. and Wigner, E. On an algebraic generalization of the quantum mechanical formalism. *Annals of Math.* (2) **35** (1934), 29–64, particularly pp. 51–54.

[**18**] Lee, H.-C. Sur le théorème de Hurwitz–Radon pour la composition des formes quadratiques. *Comment. Math. Helv.* **21** (1948), 261–269.

[**19**] Manning, H. P. *Geometry of four dimensions* (New York, 1914; Dover reprint, 1956).

[**20**] Radon, J. Lineare Scharen orthogonaler Matrizen. *Abh. Math. Sem. Univ. Hamburg* **1** (1922), 1–14.

[**21**] Semple, J. G. On complete quadrics (II). *J. London Math. Soc.* **27** (1952), 280–287.

[**22**] Serre, J.-P. *Représentations linéaires des groupes finis* (Paris, 1967).

[**23**] Severi, F. Sulla varietà che rappresenta gli spazi subordinati di data dimensione immersi in uno spazio lineare. *Annali di Mat.* (3) **24** (1915), 89–120. Reprinted in *Memorie Scelte*, 405–440.

[**24**] Stiefel, E. Richtungsfelder und Fernparallelismus in n-dimensionalen Mannigfaltigkeiten. *Comment. Math. Helv.* **8** (1936), 305–353.

[**25**] Tits, J. Sur la trialité et certains groupes qui s'en déduisent. *Publ. Math. I.H.E.S.* **2** (1959), 13–60.

[**26**] Todd, J. A. Conics in space, and their representation by points in space of nineteen dimensions. *Proc. London Math. Soc.* (2) **36** (1932), 172–206.

[**27**] Wong, Y.-C. *Isoclinic n-planes in Euclidean $2n$-space, Clifford parallels in elliptic $(2n-1)$-space, and the Hurwitz matrix equations* (American Math. Soc. Memoirs, No. 41, 1961).

INDEX

additive set, 8
algebraically parallel, 35
augmented generator, 29
autoparallel, 24
autopolar, 22
bridging construction, 66
Bridging Theorem, 69
  application in S_7, 76
  application in S_{15}, 77

central reflection, 106
chordal space, 20
Clifford parallel, 21
  α-parallel, β-parallel, 23
  in elliptic space, 4, 7
Clifford quadric, 6
Clifford regulus, 23
  α-regulus, β-regulus, 23, 122
  canonical equation, 27
  in elliptic three-space, 6
compatibility, 31
core variety, 94

directrix line, 12, 29
distance, non-Euclidean, 2
dual matrix, 63

Embedding Theorem, 36
equidistant curve, 3
equidistant surface, 4

fibrations of elliptic spaces, 5, 8, 80

generator (of regulus), 12
Grassmannian variety, 17

half-Grassmannian, 113
Hurwitz–Radon matrix equations, 52, 127
  extended form, 37

irreducible system of Clifford parallels, 36
isoclinic spaces, 6, 8

kernel of Clifford regulus, 29

linear system of Clifford parallels, 36
  affine part, 47
  degenerate, 91
  half-fixed, 52, 77
  maximal, 52
  non-singular, 48
  self-collineations, 104
  singular, 48
  space-filling, 53
  totally singular, 48

Matrix Cross-ratio Theorem, 14

natural correlation, 12
non-Euclidean geometry, 1, 3, 23
nuclear primal, 87
nuclear quadric, 89

perpendicular, 3, 4, 20
Petersen–Morley Theorem, 100

quadric in S_{2n-1}, standard equations and properties, 11
quadric-generator, 29
quadric-solid, 14

regular system, 74
regulus, 12
  Ω-regulus, 33

singular space, 47
skew reflection, 106
split regulus, 91
Study involution, 79
Study representation, 99
Study triality, 124

terminal generator, 23
terminal quadric, 29
T-representation, 82 ff.
  for S_3 and S_5, 86, 98
trivially parallel, 21

Veronesean, 18

weight of a linear system, 48